Charles O. Ofoegbu (Ed.)

Groundwater and Mineral Resources of Nigeria

International Monograph Series
on Interdisciplinary
Earth Science and Applications

Charles O. Ofoegbu (Ed.)

Groundwater and Mineral Resources of Nigeria

Friedr. Vieweg & Sohn Braunschweig / Wiesbaden

CIP-Titelaufnahme der Deutschen Bibliothek

Groundwater and mineral resources of Nigeria /
Charles O. Ofoegbu (ed.). – Braunschweig;
Wiesbaden: Vieweg, 1988
(Earth evolution sciences)
ISBN 3-528-06324-6

NE: Ofoegbu, Charles O. [Hrsg.]

Vieweg is a subsidiary company of the Bertelsmann Publishing Group.

Produced by W. Langelüddecke, Braunschweig
Printed in Germany

ISBN 3-528-06324-6

Contents

Editorial

A collection of papers on the groundwater and mineral resources of Nigeria are presented in this monograph.

The rapid growth of population, urbanisation, industrialisation and irrigated commercial agriculture and an increasing awareness of the need for adequate water supply by governments, communities and public and private institutions have contributed to the enormous demand on the surface water supply in the country. The provision of potable water has also become one of the cardinal objectives of successive governments in Nigeria over the past decade. For sustainable water supply, emphasis is now placed on the exploitation of the groundwater resources rather than the traditional dependence on surface water either from rainfall or from streams. Consequently several phases of water borehole development programs have been embarked upon by both State and Federal Governments in Nigeria in order to supplement the efforts of individuals and communities. However, in spite of the enormous financial outlays, most of the national water borehole drilling ventures have been abortive. These failures have resulted from the lack of adequate geophysical, geological and hydrogeological investigation prior to drilling and groundwater exploitation.

A collection of thematic papers dealing with the groundwater resources of Nigeria constitute the first section of this monograph. The scope of these papers include the geological appraisal of the groundwater resources, geophysical assessment of aquifer transmissivity and the siting of boreholes, hydrogeological and chemical characterisation of the groundwater resources and the pollution of aquifers in various parts of Nigeria. Also included are papers dealing with the environmental impact of surface and groundwater fluxes with particular reference to soil erosion and the development of gullies in eastern Nigeria.

The second section of this monograph addresses some aspects of the mineral resources of Nigeria. Particular attention has been focussed on these mineral deposits which although of great industrial potential in Nigeria, have not been fully exploited; these include iron ore and the non-metallic minerals. With the drastic fall in revenue from petroleum due to the low price of crude oil, and the need to diversify the mining sector of the Nigerian economy, the need to assess other mineral resources is very critical. Recently, Nigeria has realised more than ever before the importance of a viable iron and steel industry if rapid industrialization is to be realised. Several iron and steel mills were recently established by the Nigerian Government and when fully operational, these mills will require more than five million tons of iron ore annually. Studies of the geology, characterisation and the distribution of iron ore deposits in Nigeria are presented here. Virtually every part of Nigeria is thought to have deposits of one economic mineral or the other. While some measure of attention had been paid to the metallic minerals, little or no attention has been devoted to the non-metallic mineral potentials of the country. A review of the geological framework and genesis of non-metallic minerals in Nigeria is therefore presented here.

Charles O. Ofoegbu

Hydrology and Chemical Characteristics of Surface and Groundwater Resources of the Okigwi Area and Environs, Imo State, Nigeria

C. O. Okagbue
Department of Geology, University of Nigeria, Nsukka, Nigeria

Keywords: Okigwi area and environs, Water Availability, Aquifer Parameters, Water Quality

ABSTRACT

The study area of this paper lies within the southeastern Nigeria sedimentary basin and is underlain by the Asu River Group, the Eze Aku Shale, the Awgu Shale, the Nkporo Shale and the Mamu Formation. Other Formations underlying the area include the Ajali Sandstone, the Nsukka Formation, the Imo Shale Group, the Bende Ameki Group and the Benin Formation. There is high density of rivers and streams most of which have high discharges and supply the Imo River, forming the Imo River basin. The most important geologic units that bear aquifers of regional distribution are the Benin Formation, the Nanka Sands and the Ajali Sandstone Formation. The hydraulic conductivity of the aquifer units calculated from grain size analysis data varies between 0.80×10^{-2} - 10.9×10^{-2} cm/s for the Benin Formation, 8.92×10^{-3} - 7.75×10^{-2} cm/s for the Nanka Sands and 2.78×10^{-5} - 6.25×10^{-2} cm/s for the Ajali Sandstone. The transmissivity for the same aquifer units ranges between 1.18×10^{2} m2/day and 5.82×10^{2} m^2/day for the Benin Formation, 1.79×10^{2} - 5.42×10^{2} m^2/day for the Nanka Sands and 4.08×10^{-1} - 1.48×10^{2} m^2/day for the Ajali Sandstone. These values indicate aquifers of very good yield.

Most of the water samples (surface and ground water) show high concentrations of iron. The groundwaters are generally more acidic than the surface waters which show high turbidity and prohibitive concentrations of bacteriological pollution. The water resouces thus need adequate treatment for a potable utilization.

INTRODUCTION

Within the last decade or so, the accelerated growth of Nigeria in the fields of urbanisation, population, business as well as industry has led to some attention in assessment of available water resources in various parts of the country. Many workers (du Preez and Barber, 1965; Jackson, 1978; Faniran and Omorinbola, 1980; Ofodile, 1983; Egboka, 1983; Akujieze, 1984; Ogbukagu, 1984; Uma, 1984) have made significant contributions in this area. Water is important because the attainment of the goals of any society as well as the health and well-being of the population depends on a plentiful and reliable supply of this natural resource. Water forms an indispensable input into economic activities such as commerce, tourism and industry. The results of the various researches have revealed that water resources (surface and groundwater) in many parts of the country, especially the southern part, are more than adequate to meet any demand and only need development.

A complete appraisal of available water resources is often best accomplished when aspects of water quality are included. This is because in a planned water supply system, quality constraints and requirements dictate the sources of water allocated to various stages. A public water supply, though contributing greatly to the human health and well-being, can also be a vehicle for spreading disease if not properly handled. In this paper, water quality of selected water resources (surface and groundwater) in some parts of Imo State, Nigeria are reported and suggestions advanced for their healthy untilization.

GEOHYDROLOGY OF THE STUDY AREA

The study area is located between latitudes 5° 30'N and 6° 00'N and longitudes 7° 00'E and 7° 30'E and covers Okigwi and surrounding towns. It also includes towns east and south of Orlu and west of Umuahia (Fig. 1). Geologically the area lies within the southeastern Nigeria sedimentary basin and is underlain by (from oldest to youngest) the Asu River Group (Albian), the Eze Aku Shale (Turonian), the Nkporo Shale and Mamu Formation (Campano-Maestrichtian), the Ajali Sandstone Formation (Maestrichtian), the Nsukka Formation (Maestrichtian-Paleocene), the Imo Shale Group, including the Umuna Sandstone (Lower Paleocene), the Bende-Ameki Group including the Nanka Sands (middle Eocene) and the Benin Formation (Pleistocene-Oligocene) (Fig. 2).

Two major physiographic units are discernible: The Coastal Plain zone occupying a greater proportion of the southern part and the Plateau and escarpment zone occupying the north-northeastern part. The former slopes gradually from the north towards the south and southwest with slopes in the northern part steeper than those of the central and southern part which are more or less flat. The latter shows a more complicated topography of isolated hills that are separated by steep valleys. The topography reaches an elevation of 400 m (1300') north of Okigwi with the main scarp forming an arc that swings to the east from Okigwi towards Afikpo.

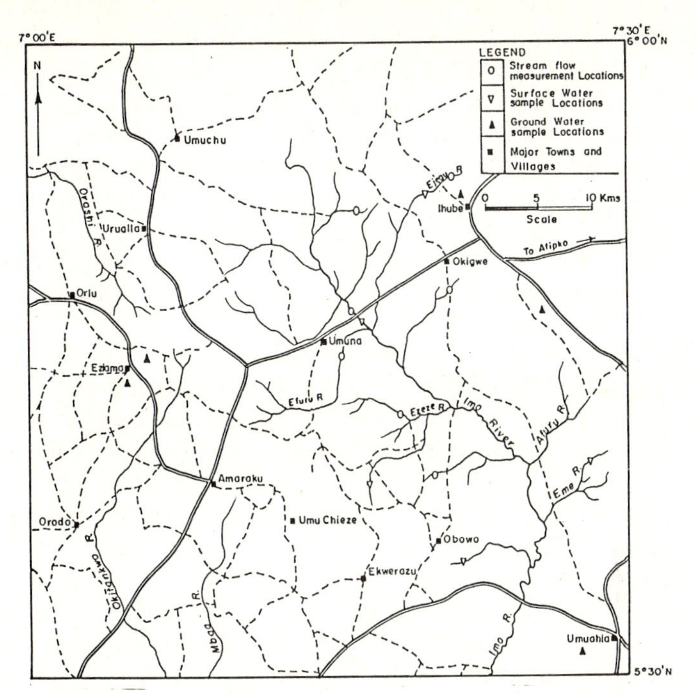

Fig. 1. Map showing sample/measurement locations in the study area

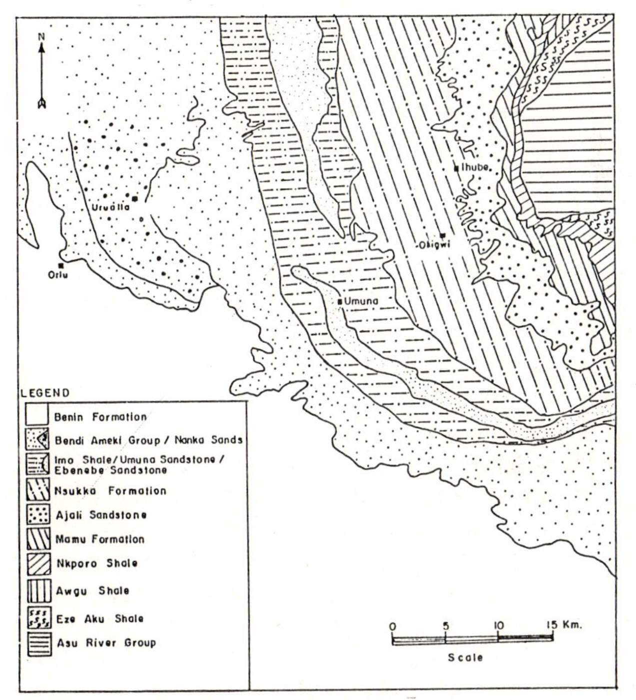

Fig. 2. Geological map of study area

The main rivers from which surface water schemes could be developed flow generally north to south and include the Imo River, the Okitankwo River, the Mbaa River, the Efuru River, the Ezeze River, the Nfuru River and the Eme River. Most of these rivers feed the Imo River and form the Imo River basin, the largest of the drainage basins in Imo State. The river originates outside the coastal plains in the shale lowlands north west of Okigwi and flows southeastwards receiving most of the tributary rivers, but swings southwards in the southern part of the study area where it receives some other major tributaries including the Oramirukwa and Otamiri Rivers draining the area around Owerri. Most of these streams discharge significant volumes of water as summarized in Table 1.

GENERAL HYDROSTRATIGRAPHY AND AQUIFER CHARACTERISTICS

The most important geologic units that bear aquifers of reginal distribution are the Benin Formation, the Bende Ameki Group, and the Ajali Sandstone Formation. Aquiferous layers of local importance are found in the Nsukka Formation and in the river bed alluvium. Local and mostly pierced sandy aquifers are also found within the thick shale sequences of the Imo, Nkporo, Awgu and Eze Aku Shale Formations.

The Benin Formation whose thickness averages 1000 m in the study area is built of thick unconsolidated sands interfingered with clay bands, lenses and stringers. The sand is mostly coarse grained, pebbly and poorly sorted. The sand and clay intercalations constitute a system of aquifers separated by aquitards. The aquifer-aquitards units form a multi-aquifer system. It is estimated from lithologic logs that the unconfined aquifer zone varies in thickness from 30 m to 40 m. Beds of sand of over 40 m thick have been recorded at Umuna-Orlu for a middle confinded to semiconfined aquifer while drill records, although shallow, indicate presence of a lower confined aquifer whose thickness cannot yet be established due to relatively shallow penetration of the existing wells.

The hydraulic conductivity (K) and transmissivity (T) of the aquifers are high. The hydraulic conductivity of the middle aquifer is placed in the range 0.72×10^{-2} - 24×10^{-2} cm/s (from pumping test data) and 0.8×10^{-2} - 10.9×10^{-2} cm/s (from grain size and statistical data). The transmissivity is placed in the range 1.18×10^{2} m^2/day - 5.82×10^{3} m^2/day (from pumping test). These values indicate an aquifer of very good yield (clean sand) (Freeze and Cherry, 1979). Although a quantitative evaluation of the aquifer parameters of the unconfined aquifer is difficult (since no special pumping tests have been performed for this aquifer unit in the region), a range of coefficients can be assumed based on general data as well as on data from adjacent regions. The hydraulic conductivity which is a function of effective grain size is estimated to range between 3.53×10^{-2} cm/s and 1.2×10^{-2} cm/s depending on the percentage of coarseness of the sand component while the transmissivity may reach 1.03×10^{3} m^2/day (Uma, 1984).

Table 1 Summary of River/Stream Flow Data

River/Stream name	Measuring Location	Month/s of measurement	Flow measurement m^3/Sec
Imo	Umuokwara	April, July, Oct.	11.5; 99.5; 62.0
"	Ndimoko	"	1.1; 12.1; 6.49
"	Umuna Bridge	"	2.09; 21.4; 35.6
Ejegu	Aku-Ihube	June	0.95
Efuru	Bridge to Umuelemai	"	1.5
"	Obiohoro-Osu	February	0.7
Onuezeala	Ugiri	"	0.3
Ezeze	Nzerem	"	0.07
Onuiyin	Okwuohia	"	0.09
Asa	Umueze	"	0.03

* Source, Imo River Basin authority.

Table 2 Aquifer Parameters of Benin Formation, Nanka Sands and Ajali Sandstone Formation.

Formation	Range of Aquifer Parameter	Aquifer Unit	Parameter Calculated	Basis of value
Benin Formation	0.72×10^{-2}–24×10^{-2} cm/s	Middle Confined	K	From pumping test data
	0.80×10^{-2}–10.9×10^{-2} cm/s		K	From grain size data
	1.18×10^{2}–5.82×10^{3} m^2/day		T	On basis of screen length using Logan, 1964 Giusti, 1978 equations
	1.2×10^{-2}–3.53×10^{-2} cm/s	Unconfined	K	From grain size data
	6.15×10^{2}–1.03×10^{3} m^2/day	"	T	On basis of aquifer thickness and K value.
Nanka Sands	8.92×10^{-3}–$7.75 \times$ cm/s	Lower confined	K	From grain size data
	5.73×10^{2}–1.25×10^{3} m^2/day	"	T	On basis of aquifer thickness and K value
	1.79×10^{2}–5.42×10^{2}	"	T	From pumping test
Ajali Sandstone	8.32×10^{-3}–6.25×10^{-2} cm/s		K	From grain size data
	2.78×10^{-5}–6.67×10^{-3} cm/s		K	From pumping test
	4.08×10^{-1}–1.48×10^{2} m^2/day		T	From pumping test

The Ameki Formation which includes the Nanka Sands is composed of a lower unit of alternating sandstone, siltstone, and mudstone and an upper unit of mainly sands - the Nanka Sands. The sand which is medium to coarse grained is highly porous and permeable. It reaches a maximum thickness of 300 m. Akujieze (1984) has identified four sand subunits in the Nanka Sands and has estimated aquifer thicknesses of up to 42 m from this sand. The aquifers occur in confined, semiconfined and unconfined conditions at different places. Test yields varying from 55 m^3/hr (14531 gph) to 66 m^3/hr (17437 gph) have been recorded of the fourth sand subunit, thus showing that this subunit is a good aquifer.

Hydraulic conductivity estimated from statistical analysis of grain size parameters (Hazen 1911; Carman-Kozeny, 1956; Harleman, et al., 1963; Masch and Denny, 1966) give values ranging between 8.92 x 10^{-3} to 7.75 x 10^{-2} cm/s. These values are very close to the values calculated for the Benin Formation aquifer. Hence the Nanka Sand aquifer is also of very good yield. The transmissivity estimated from statistical values ranges between 5.73 x 10^2 and 1.25 x 10^3 m^2/day while that calculated from step-drawtown test ranges between 1.79 x 10^2 and 5.42 x 10^2 m^2/day. The transmissivity of the Benin Formation aquifer is higher than that of the Nanka Sands aquifer. This is expected considering that the Benin Formation aquifer is thicker than the Nanka Sands aquifer.

The Ajali Sandstone consists of friable, coarse grained and poorly sorted sandstone. This formation also has high hydraulic conductivity. The values calculated from grain size data using several methods (Hazen 1911; Harleman et al., 1963, Masch and Denny, 1966) range between 8.32 x 10^{-3} and 6.25 x 10^{-2} cm/s while those calculated from pumping test analyses range between 2.78 x 10^{-5} and 6.67 x 10^{-3} cm/s. The transmissivity ranges between 4.08 x 10^{-1} m^2/day and 1.48 x 10^2 m^2/day (from pumping test analyses). In general the Ajali Sandstone, though a good aquifer, is not as prolific as the Benin Formation of the Nanka Sands. Table 2 shows several parameters of each of the three formation aquifers.

Other aquiferous layers of local importance include the sandstone layers of the Nsukka Formation, the sand intercalations within the Imo, Nkporo, Awgu and Eze Aku Shale Formations and also some river-bed alluvium. These aquifers occur locally; thus their importance from the regional hydrogeological point of view is limited. They are, however, useful when local problems of water supply are considered. In general, their location may demand a detailed geological and geophysical survey, probably accompanied by exploratory/test drilling.

WATER QUALITY TEST RESULTS

Water properties which are important from the public health point of view include physical as well as chemical properties. Physical properties include taste, odour, turbidity, and colour while chemical properties include hardness, iron, nitrate, chloride and other ions content. Corrosion and/or precipitation-producing elements

Table 3 Results of Surface water quality Analysis (values in mg/l)

River/Stream	Location	Color (Hazen units)	Turbidity	Conductivity µ mhos/cm	pH	Magnesium hardness	Calcium hardness as $CaCO_3$	Total Hardness as $CaCO_3$ (mg/l)	Ca^{2+}	Mg^{2+}	Total iron as Fe	Chloride as NaCl	Cl	NO_3	T.S at 103°C	SiO_2
Nfuru	Umuobiala	50	nd	22.0	8.2	4.0	2.0	6	nd	nd	1.5	nd	2.1	0	18	15
Imo	Umuna	100	nd	24.0	7.5	3.0	3.0	6	nd	nd	2.5	nd	1.4	0.06	78	25
Onuezeala	Ugiri	5	nd	10.0	8.5	3.0	2.0	5	0.6	0.8	0.22	10.6	6.4	0.03	32	8
Onuiyi	Etiti	5	nd	10.3	6.8	2.5	1.5	4	0.6	0.8	0.36	4.1	2.5	0	38	8
Duru-Owerre	Obollo	5	nd	9.5	7.9	2.5	3.5	6.0	1.4	0.8	2.0	5.8	3.5	0	40	6
Ezeala Anyanwu	Mbano	90	nd	10.2	7.8	2.0	2.0	4.0	0.8	0.6	2.98	6.4	3.9	0	46	10
Ejegu	Ihube	40	16	11.0	6.1	5.0	2.0	7.0	nd	nd	nd	nd	0	0.11	90	12
Okitakwo		75	23	15.5	5.9	5.0	2.0	7.0	nd	nd	0	nd	0	3.8	80	9
Iyi Ucha	Okigwi Urban	5	nd	nd	6.4	69	37	106	nd	nd	nd	nd	nd	9.7		5
Ibiyi	Isuikwuato	nd	nd	20	6.3	15	5	20	2	nd	nd	nd	nd	13.2		14

nd = not determined

which may cause heavy damage to water supply systems, toxic elements as well as bacterial pollution which may be harmful to human beings are also important.

Table 3 shows analyses results of randomly sampled surface waters representing many rivers and streams which can be considered as sources of water supply. The colour varies from 5 Hazen units in clear and colourless streams to 100 Hazen units in turbid or cloudy streams and rivers. The pH of the water samples varies from 5.9 to 8.5 indicating that some of the streams are weakly acidic while others are slightly alkaline. Experience has shown that water having pH below 6 may be corrosive to pipes and equipment.

The amounts of different elements present in milligrains per litre (mg/l) in the tested samples are also shown in Table 3. The hardness due to the presence of calcium and magnesium ions ranges from 1.5 - 3.0 mg/l in calcium 50 2.0 - 5.0 mg/l in magnesium. Total hardness as $CaCo_3$ varies from 4.0 to 7.0 mg/l while total dissolved solids vary from 18 to 90 mg/l. Ezeala Anyanwu River has the highest value of iron content of 2.98 mg/l. Chlorides as NaCl ranges from 4.1 mg/l in Onuiyi River to 10.6 mg/l in Onuezeala River.

Table 4 gives the analyses results of a number of ground water samples from selected locations. In terms of total dissolved solids (TDS), the waters exhibit very low concentrations of chloride (Cl^-) and other ions except iron (Fe). The levels of alkalinity (bicarbonate anion) and calcium are also very low, a fact that can cause aggressiveness or corrosive effects from the waters. Obviously the analyses results show that there are undetermined constituents in the waters analysed as the sum of the cations and the sum of the anions when expressed in milliequivalents per litre are not equal.

DISCUSSION

From all indications, the surface waters are much more polluted than the ground water. The amount of sodium chloride (NaCl) recorded in the surface waters (4.1 - 10.6 mg/l) is high and suggests significant interaction between the waters and such substances as clay minerals, chloride and mica which are capable of yielding Na and Ca ions in solution. The iron contents in both surface and ground water samples are high (0.22 - 2.98 mg/l in surface waters and 0.1 - 0.7 mg/l in ground water). When compared with the international standard for drinking water (Table 5), many of the samples (surface and ground water) have iron content in excess of the allowable upper limit. The iron is thought to be principally derived from the overburden of red earths characteristically overlying most of the formations in the study area. It is also possible that part of the iron emanates from chemical weathering of minerals such as pyrite and marcasite. Pyrite, and possibly marcasite, occurs in intervening shale beds of most formations. Most villagers who draw water from boreholes complain of objectionable taste from the ground water. It is suspected that such taste results from

Table 4: Properties of selected ground water samples (values in mg/l)

Source of Water	pH	Calcium hardness	Total hardness as $CaCo_3$	Alkalinity as $CaCO_3$	Cl	NO_3	F	Na	Total Fe	TDS
Umuna-Orlu	6.9	9.0	12.0	33.0	3.0	0.50	0.0	12.0	nd	44
Ihube-Okigwi	6.0	1.5	2.0	2.5	1.1	1.00	0.1	0.5	0.1	6
Umuahia	6.0	7.0	7.5	7.5	1.6	1.5	0.1	1.2	0.4	12
Eziama-Orlu	6.0	6.5	7.0	12.5	1.0	1.0	0.1	0.6	1.1	15
Amawo-Isikwu-ato	6.0	5.5	6.0	7.5	1.0	1.0	0.1	0.4	0.7	10

Table 5 Standards and Criteria for drinking water (mg/l)

Substance or property	Public Health Service 1962		EPA interim 1975	WHO, 1963		National Acad. Sci National Acad. Eng. 1972
	Desirable max. limit	Absolute max. limit		Max. acceptable	Max. allowable	
Ammomum nitrogen	-	-	-	-	-	0.5
Arsenic	0.01	0.05	0.05	-	0.05	0.1
Cadinuim	-	0.01	0.01	-	0.01	0.01
Barium	-	1	1	-	1	1
Calcium	-	-	-	75	200	
Chloride	250	-	-	200	600	250
Chromium (hexavalent)	-	0.05	0.05		0.05	0.05
Color (Pt-Co Units)	-	-	-	5	50	75
Copper	1	-	-	1	1.5	1
Cyanide	0.01	0.2			0.2	0.2
Fluoride	0.6-0.9	0.8-1.7	1.4-2.4	-	-	1.4-2.4
Iron (Fe^{2+})	0.3	-	-	0.3	1	0.3
Lead	-	0.05	0.05	-	0.05	0.05
Magnesium	-	-	-	50	150	-
Magnesium and Sodium Sulphates	-	-	-	500	1000	-
Manganese (Mn^+)	0.05	-	-	0.1	0.5	0.05
Mercury	-	-	0.002	-	-	0.002
Nitrate nitrogen	10	-	10	-	-	10
Nitrite nitrogen	-	-	-	-	-	1
pH (Units)	-	-	-	7-8.5	-	5-9
Phenolic Compels (as phenol)	0.001	-	-	0.001	0.002	0.0000-.01
Selenium	-	0.01	0.01	-	0.01	0.01
Silver	-	0.05	0.05	-	-	-
Sulphate	250	-	-	200	400	250
Total dissolved Solids	500	-	-	500	1500	-
Zinc	5	-	-	5	15	5

high concentrations of iron. More than 0.1 mg/l of iron in groundwater imparts objectionable tastes and colours to food and drinks (Durfer and Baker, 1964).

The range (6 - 44 mg/l) of total dissolved solids (TDS) in the tested ground water samples almost falls within that of surface water which varies from 18 to 90 mg/l. Both surface and ground water samples are low in TDS. WHO (1971) and Johnson (1975) note that water containing less than 500 mg/l of dissolved solids is generally satisfactory for domestic use and for many industrial purpose.

All the tested samples (surface and ground water) are characteristically soft (range of total hardness as $CaCo_3$ is 4 - 6 mg/l for surface water and 2 - 12 mg/l for ground water) following Linsley and Franzini's (1979) classification of water hardness. The waters are thus suitable for laundry as well as for commercial boiler operations. However, the high iron content would stain laundry.

Preliminary bacteriological tests reveal that most of the tested samples show total coli or Escherichia coli in prohibitive concentrations (80 - 900 per 100 ml), indicating bacterial pollution of the water sources particularly in surface waters. The WHO standard requires that (i) the water shall contain no bacteria coli type 1 (faecal coli per 100 ml of all water samples) and (ii) the water shall contain no presumptive coliform organisms per 100 ml in 98 % of all samples. The waters in their present state are thus unsatisfactory for drinking as they are contaminated by faecal and other biological materials. Some of the concentrations of nitrate (NO_3) recorded from the samples may also have come from animal excrement.

CONCLUSIONS AND RECOMMENDATIONS

The development of water supply schemes which will have safe and sufficient quantities of water of adequate quality for all the rural communities in Okigwi area is a matter of vital importance. At present the existing water supply facilities are very limited and cover only a small part of the study area. As this study has shown, water resources (both surface and ground water) are plentiful but lack the quality for human consumption.

From the analyses results, the sources of water in the area can be characterised by a number of specific deficiencies such as: (i) High aggressiveness (corrosivity) due to low alkalinity and calcium hardness and/or low pH values observed mostly in the ground water.
(ii) Turbidity (above permissible levels) due to the presence of suspended solids, characteristic of all surface waters
(iii) Concentrations of certain ions (such as Fe) above permissible levels in both surface and groundwater
(iv) Total coli or Escherichia coli in prohibitive concentrations, showing bacterial pollution of the water sources particularly in surface waters.

When the Government is ready to develop a water scheme for the study area, treatment will be required to eliminate the impurities. In each case the quality of the water at the source will dictate the required treatment level and definite steps to be employed. However, because of the similarity of qualities in almost all ground water samples as a group, as well as the river/stream samples as a group, a general method of treatment could be used for each water resource according to its grouping even if the available analyses are not complete. The following recommendations are therefore proposed on a group basis:

Surface Waters: The method for treating stream water into drinking water should be considered as one which on the one hand guarantees an optimal treatment with minimal usage of chemicals, and on the other hand permits flexible adaptation to a possibly changing raw water quality. As the concentration of suspended solids (whether of organic nature or inorganic silt) fluctuates as a result of weather conditions in every season, a flocculation process will fulfill the requirement for flexible adaptation regarding the suspended solids. Thus the following procedure is proposed:
(i) Flocculation by addition of alum or aluminum sulphate
(ii) Addition of hydrated lime to adjust pH value
(iii) Flocculation aided by polyelectrolyte if necessary
(iv) Clarification by sedimentation
(v) Filtration in either rapid gravity or pressure filters
(vi) Possible adjustment of pH value of clear water after addition of lime.
(vii) Disinfection of clear water by chlorination.
The flocculation process includes basically the flocculation and sedimentation of the suspended colloidal and the colored soluble pollution as well as the soluble iron. Filtration of the water will be necessary for further purification. The addition of lime, which is a required step when flocculation with alum or aluminum sulphate is applied, acts also a a neutralizer of the undesirable acidity which is characteristic of the raw water.

Groundwater: Groundwater pumped from boreholes will have to be neutralized in order to eliminate its aggressive nature. This can be achieved by the addition of either lime in the form of carbonate or bicarbonate, thus increasing both the alkalinity and the calcium hardness, or by adding hydrated lime thus raising the pH and increasing the calcium hardness. Such neutralization methods are aimed at getting a favourable saturation pH and Index (Langelier Index) which is an indication of the aggressive or lime precipitation capacity of the water. Where necessary, iron will have to be removed from the water either by aeration or by lime treatment followed by a separation step, in which the oxidized ferric hydroxide sediments are removed from the water. Disinfection by chlorination is also recommended as a precautionary measure.

ACKNOWLEDGEMENT

The author expresses gratitude to Mr. C.J. Uzowulu of the Anambra State Water Board for the chemical tests.

REFERENCES

Akujieze, C.N., 1984: The Hydrogeology of the Area (Nanka Sand) South of the Upper Sector of the Idemili Drainage Basin. Unpublished M. Sc. thesis, Dept. of Geology, University of Nigeria, Nsukka.

Carman, P.C., 1956: Flow of Gases through porous media, Butterworths, London.

Durfer, C.N. and E. Baker, 1964: Prinzipal Chemical Constituents in Groundwater - Their sources, concentrations and Effect on usubility. U.S. Geol. Survey water-supply paper 1812.

Egboka, B.C.E, 1983: Analysis of the groundwater resources of Nsukka area and environs, Anambra State, Nigeria. Nigeria. Jour. of Min. Geol. vol. 20 Nos. 1 & 2, pp. 1 - 16.

Faniran, A. and E.O. Omorinbola, 1980: Evaluating the shallow groundwater reserves in Basement Complex areas: A case study of Southwestern Nigeria. Nigerian Jour. Min. Geol. vol. 17, No. 1, pp. 65 - 79.

Freeze, R.A. and J.A. Cherry, 1979: Groundwater, Prentice-Hall Inc., Englewood Cliffs, New Jersey.

Giusti, E.V., 1978: Hydrogeology of the Karst of Puerto Rico. U.S. Geol. Survey Prof. Paper 1012.

Harleman, D.R.F., P.F. Mehlhorn, and R.R. Rumer, 1963: Dispersion -Permeability correlation in porous media. Jour. Hydraul. Div., Amer. Soc. Civil Eng., vol. 89 (HY 2), pp. 67 - 85.

Hazen, A., 1911: Discussion of "Dams on sand foundations" by A.C. Koenig. Trans. Amer. Soc. Civil Eng., vol. 73, pp. 199 - 203.

Jackson, J.O., 1978: Geotechnical concepts for the Lagos water supply expansion programme. Nigerian Jour. Min. Geol. vol. 15, No. 2, pp. 72 - 77.

Johnson, U.O.P. 1975: Groundwater and Wells. Johnson Division. St. Paul. Minnesota, U.S.A.

Linsley, K. and J.B. Franzini, 1979: Water Resources Engineering. McGraw-Hill, New York.

Logan, J., 1964: Estimating Transmissivity from Routine Production Tests of water wells. Groundwater, vol. 2, No. 1, pp. 35 - 37.

Masch, F.D., and K.J. Denny, 1966: Grain-size Distribution and its effect on the Permeability of Unconsolidated sands, Water Resources Research, vol. 2, pp. 665 - 677.

National Academy of Sciences and National Academy of Engineering, 1972: Report prepared by Committee of Water Quality Criteria at request of U.S. Environmental Protection Agency, Washington, D.C.

Ofodile, M.E., 1983: The occurrence and exploitation of groundwater in Nigerian Basement rocks. Nigerian Jour. Min. Geol. vol. 20 Nos 1 & 2, pp. 131 - 146.

Ogbukagu, IK., 1984: Hydrology and Chemistry of surface and groundwater resources of the Aguata area, SE Nigeria. Jour. African Earth Sciences, vol. 2, No. 2, ppa. 109 - 117.

Preez, J.W. du and D.F.M. Barber, 1965: The distribution and chemical quality of groundwater in northern Nigeria. Bull. geol. Surv. Nigeria No. 36, pp. 1 - 93.

Public Health Service, 1962: Drinking water standards, U.S. Dept. of Health, Education and Welfare, Washington, D.C.

Todd, D.K., 1980: Groundwater Hydrology (2nd ed.) John Wiley and sons, New York.

Uma, K.O., 1984: Water resources potentials of Owerri and its environs. Unpublished M.Sc. thesis, Dept. of Geology, University of Nigeria, Nsukka.

World Health Organisation, 1971: International Standards for drinking water, Geneva.

Aquifer Transmissivity from Electrical Sounding Data: The case of Ajali Sandstone Aquifers, Southeast of Enugu, Nigeria

K. Mosto Onuoha and F. C. C. Mbazi
Department of Geology, University of Nigeria, Nsukka, Nigeria

Keywords: Geoelectric sounding, Hydraulic conductivity and Transmissivity, Water quality, Anambra Basin, Water table

ABSTRACT

The study area lies south-west of Enugu and falls in the Udi and Ezeagu Local Government areas of Anambra State. Surface water is very uncommon in this area and infiltration of rain water into the subsurface is very rapid due to the very porous and unconsolidated nature of the underlying Ajali Formation. Twenty seven vertical electrical soundings were conducted with a maximum current-electrode separation of 1.20 km, employing the Schlumberger electrode configuration. The interpretation of the sounding data revealed three distinct geoelectric layers overlying highly resistive strata. Five of the soundings were conducted beside boreholes located in the project area. From the data, the variations of depth to the water table, the longitudinal conductance, transverse resistance, and thickness of the aquiferous horizon across the study area have been established. These have led to inferences about aquifer hydraulic conductivities and transmissivities. A better idea of the groundwater potentials of the area has also emerged from this study.

Although the area appears homogenous from surface geological mapping, two distinct zones with different hydraulic properties and water qualities have been identified. These are the Nachi-Amokwe area on the one hand, and the Obeleagu-Obioma - Ninth Mile Corner areas on the other. Hydrochemical investigations would be required to confirm and establish the degree of difference between the aquifer system under these areas. The geophysical data have also led to the precise delineation of the groundwater divide between the Anambra and Cross River drainage system in the study area.

INTRODUCTION

In Udi and Ezeagu Local Government areas of Anambra State, several boreholes have been drilled by the State Water Board and by the Federal Government under its last 'National Borehole Programme'. Some of these boreholes were dry or yielded very little water. Part of the problem was due to the fact that no proper hydrogeological or geophysical investigations were conducted before the sites for the drilling of the boreholes were selected. Water supply problems are not new to many areas of Anambra State, but the situation is very bad in some localities, like in the area of this investigation.

The present project was undertaken to help in delineating sites for the drilling of productive water boreholes in the area, and to study the hydrogeophysical properties of the Ajali Sandstone, which is the geological formation that underlies the area. The prolific nature of the Ajali Sandstone as an aquifer has been known for a long time, but the evolution of a feasible water resources management programme requires input from geological, geophysical and hydrogeological surveys, including test drilling, pump testing and water analyses. The present project is also designed to contribute to the overall management programme which will culminate in the future in a sophisticated mathematical model for managing the water resources of the Anambra basin.

PHYSIOGRAPHY, GEOLOGY AND HYDROGEOLOGY

Figure 1 shows the project area with the access roads and the locations of the sounding stations. The major communities found within the area include Nsude, Imezi Owa, Umana, Abia Udi, Obioma, Amokwe and Nachi. The old Enugu-Onitsha road and the new Enugu-Onitsha expressway pass through the area. A network of motorable and track routes make access to most parts of the area possible. Figure 2 shows that the survey area is situated between the Udi Hills to the east and the Anambra lowlands to the west, thus forming part of the eastern most limits of the Anambra basin. The eastern part of the area has higher relief (about 460 m) whereas the western and south-western portions have an average elevation of 215-240 m. The area is completely devoid of streams or rivers (seasonal or perennial), though numerous streams rise and flow eastwards from the Enugu escarpments (Fig. 2).

The paucity of streams maybe due to the high percolation rate of runoffs in the porous underlying Ajali Formation. The Ajali Sandstone (Maestrichtian) is friable, poorly sorted, poorly cemented, typically white to iron-stained and very often cross-bedded. In the study area, fresh outcrops were mapped only in a gulley at Nsude. In nearly all the other locations, outcrops were severely weathered into loose and dry top soil. The overburden in most places is lateritic, and between Nsude, Obioma and Ogwuafia Owa the iron content of the laterites have been found to be quite high (25 - 50 %; Hazell, 1955). The Ajali Sandstone is underlain by the Mamu Formation

Fig. 1 Map of the Area Showing Access Roads and Sounding Stations.

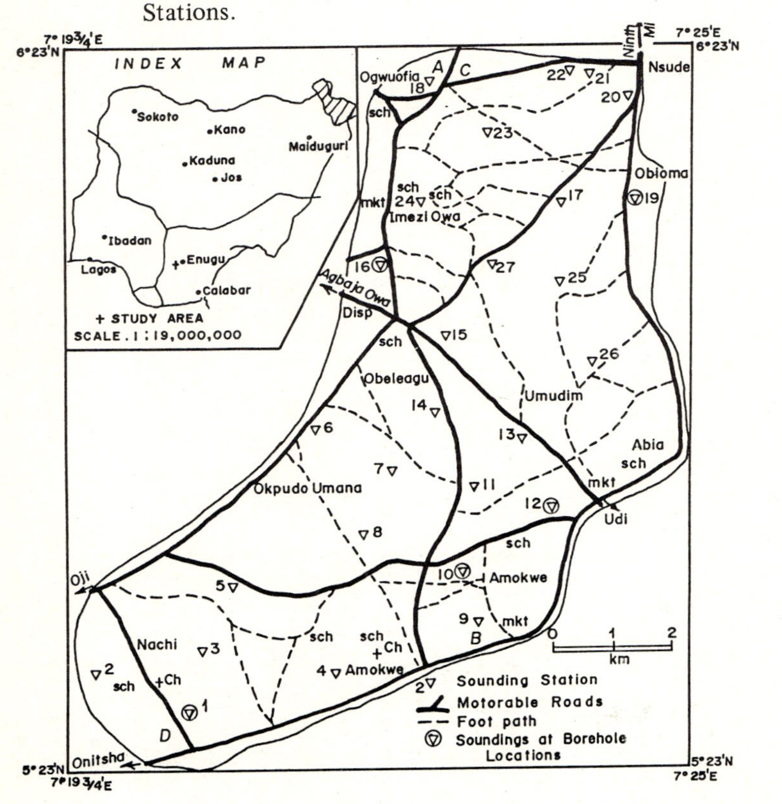

Fig. 2 Physiography of the Study Area.

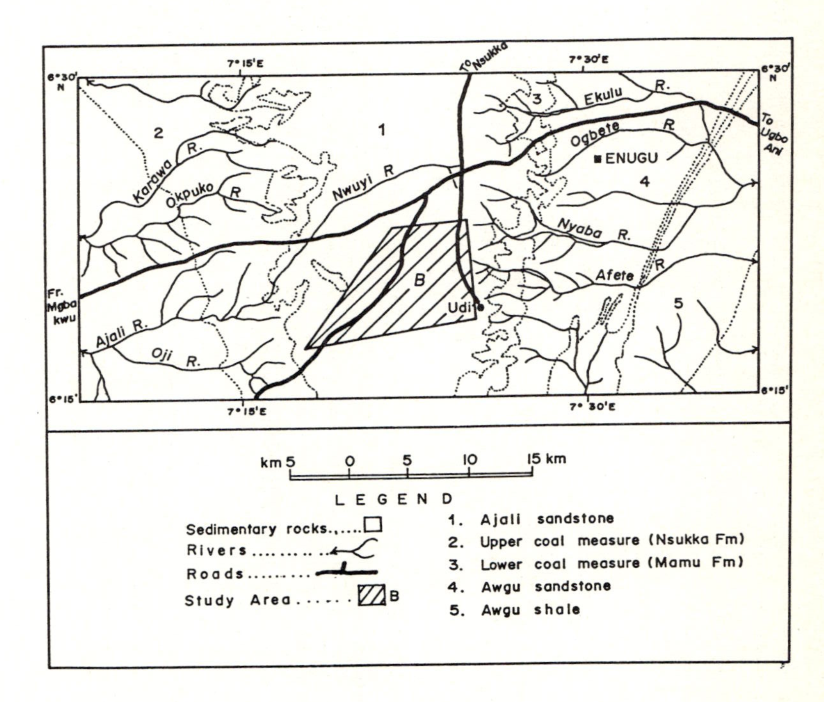

(Lower Coal Measures), and overlain by the Nsukka Formation (Upper Coal Measures). These formations however, do not outcrop in the study area.

The hydrologic condition favouring aquifer formation in the area is provided by the lateritic red earth and the weathered top of the Ajali Sandstone, and by the clay/shale members of the Mamu Formation that immediately and conformably underlie the permeable and porous Ajali Formation. This structure is instrumental to water storage and the saturation of the deeper layers of the Ajali Formation. Water table outcrops abound at the boundary between the Mamu Formation and the Ajali Sandstone on the Enugu escarpment - off the study area. These outcrops are the major sources and recharge centres of most streams that flow eastwards around Enugu and Awkunanaw.

DATA COLLECTION AND INTERPRETATION

The locations of the sounding stations are shown on Figure 1. Altogether a total of twenty seven Schlumberger soundings were made in the study area to a maximum current-electrode separation (AB) of 1.20 km. The instrument used was the ABEM SAS 300 Terrameter, a digital, averaging instrument for DC resistivity work. Five soundings were made at the sites of existing boreholes for comparative purposes to explore the interaction of subsurface materials. Existing boreholes beside which soundings were conducted are located near the Udi market, and at Obioma, Amokwe, Nachi and Obeleagu Umana. The respective vertical electrical sounding (VES) stations are 12, 19, 10, 1 and 16 at those borehole locations. The traverses were oriented nort-south and the half current-electrode spearation (AB/2) was increased in logarithmic steps. The interpretation was by the conventional partial curve-matching and the use of auxiliary point diagrams (Zhody, 1965; 1974).

Most of the sounding curves obtained were of the QH-type ($\rho_1 > \rho_2 > \rho_3 < \rho_4$) representing the presence of four geoelectric layers. The terminal branches of many of the QH curves rose steeply into positive slopes that made angles of 45° with the abscissa. It is well known that such a steep rise in a sounding curve is a reflection of highly resistive sedimentary or basement rocks at depth. The total transverse resistance T, and the total longitudinal conductance S were determined for each sounding station from the interpreted data.

$$T = \sum_{i=1}^{n} h_i \rho_i \qquad i = 1, 2, 3, \ldots n \tag{1}$$

$$S = \sum_{i=1}^{n} h_i/\rho_i \qquad i = 1, 2, 3, \ldots n \tag{2}$$

Where h_i and ρ_i denote the thickness and resistivity of the i-th layers respectively. It has been demonstrated in the past, that T and S are powerful autonomous interpretational aides in groundwater surveys (Zhody, 1974; Henriet, 1976). Aside from defining the aquifer geometry (which the resistivity soundings enable us to do), we can make inferences on transmissivity and aquifer storage on the basis of T and S values, and their distribution in space across the survey area. More details about the sounding curves and their mode of interpretation have been given by Mbazi (1985).

DATA PROCESSING AND RESULTS

In processing the data, the following steps were taken:
a) interpretation of the sounding data along profiles AB, and CD, so as to provide cross-sections;
b) development of an isopach map of the aquiferous layer from the data;
c) development of maps of the aquifer longitudinal conductance, transmissivity and the variation across the area of other diagnostic parameters (e.g. hydraulic conductivity-electrical conductivity product);
d) identification of prospective zones for the drilling of productive boreholes in the area.

Interpretative Cross-Sections Along Sounding Stations

The results of the soundings together with geological data were used to construct two interpretative sections across the study area. The positions of the cross-sections AB and CD can be ascertained from Figure 1. The points A and C are located in the north-eastern corner of the survey area near Ogwuafia, and are close to VES station 18. Point B is located in the southern part close to VES station 9, which is about 1.5 km east of the Girls' Secondary School, Amokwe. D is located in the south-eastern corner near VES station 1, which is close to the State Water Corporation Borehole at Nachi. Aka and Onuoha (1983) have reported the results of electrical resistivity soundings conducted in the Ninth Mile Corner and its environs (area A on Fig. 2), including the vicinity of the State Water Corporation 12-well field at the Ninth Mile. This area is geologically similar to the sutdy area and is a few kilometers to the north (Fig. 2). Combining the interpreted sounding results with information from the lithologic logs of the existing boreholes, we arrive at the conclusion that the first two geoelectric layers correspond to the ferruginous, lateritic overburden. The upper part of this overburden contains ironstones and laterites, while the lower part is made up of red, coarse sand that is ferruginous. The third geoelectric layer, whose composition according to the lithologic logs is mainly of white to yellow, medium to fine-grained sands has been identified as the water-bearing layer, whose resistivity varies from place to place, but is within the range 60-700 ohm m.

The interpretative sections are shown on Figures 3 and 4. The aquiferous layer thickens towards Nachi and Udi and is thin in the area between Imezi Owa and

Fig. 3 Interpretative Cross – Section Across Profile AB

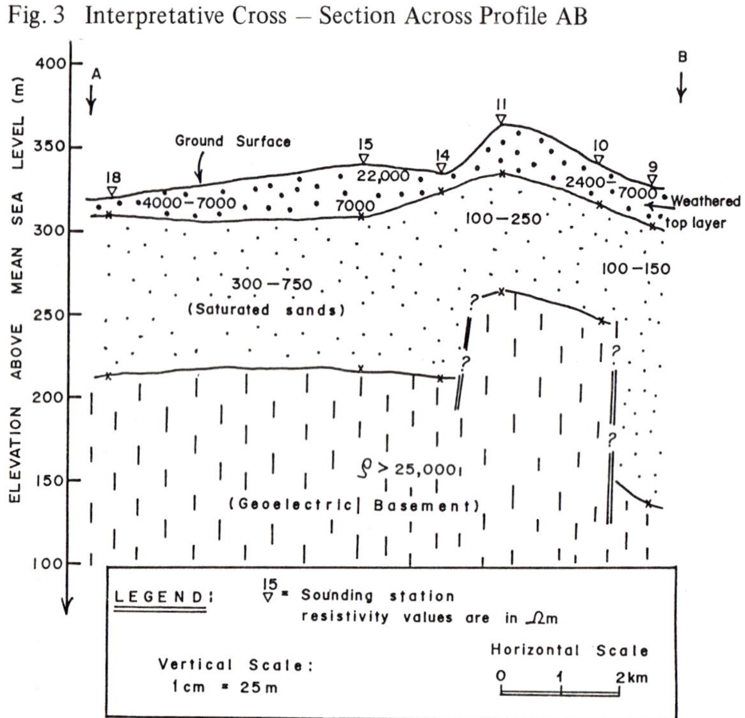

Fig. 4 Interpretative Cross – Section Across Profile CD

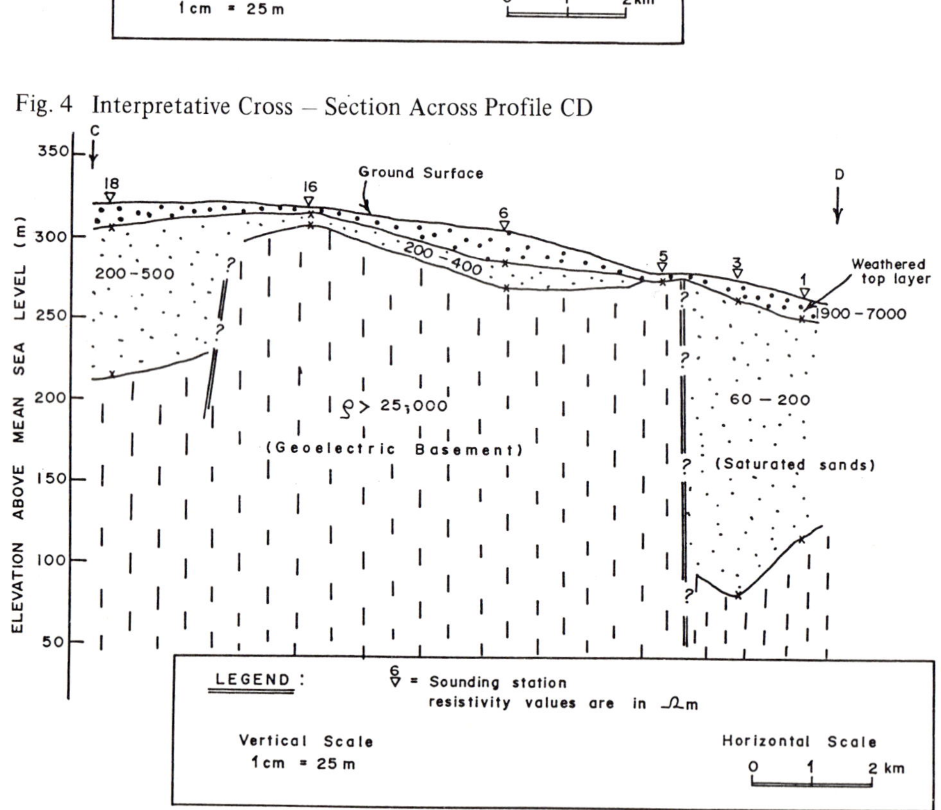

Okpudo Umana. This layer is virtually absent at the site of VES station 5, south-east of Okpudo Umana (Fig. 4). Some inferences have also been made about the depth of the weathered top soil and lateritic overburden. Two zones with thick overburden can be discerned. The first stretches from Obioma towards Nsude and the Ninth Mile Corner, and the thickness is in the range 19 - 22 m. The second zone is in the Obeleagu-Amokwe-Udi triangle to the south, and the overburden thickness is in the range 20 - 29 m. Figures 3 and 4 also reveal indications of subsurface faulting in the highly resistive 'geoelectric basement', but further geophysical work will be necessary to precisely locate such faults.

Depth to the Water Table and Isopach Map of the Aquiferous Layer

Depth to the water table has been deduced from the sounding results and the indications are that the water table is shallow in the Umana-Obeleagu-Owa area and much deeper in the Amokwe-Abia Udi-Umudim area. There is a correlation between depth to the water table and the thickness of the weathered red earth and lateritic overburden. Deeper water tables are found where thick layers of the weathered zone occur. the top of the water table also appears to mirror the surface topography as is evident from Figures 3 and 4. An isopach map (Fig. 5) of the aquiferous layer has also been prepared from the interpreted data. This figure could definitely be of great help in future water drilling programmes in the area. The unsteady yield of the borehole near Obeleagu Umana, and the abortive holes drilled in parts of Owa, just off the study area maybe partly due to the thinness of the aquifer in these areas. It is evident from the above diagrams, that well yields are likely to be better in the southern part of the study area (e.g. around Nachi, Amokwe and to the west of Obioma).

Aquifer Transverse Resistance, Longitudinal Conductance and Transmissivity

Niswas and Singhal (1981) have established an analytical relationship between aquifer transmissivity and transverse resistance on the one hand, and between transmissivity and aquifer longitudinal conductance on the other. From Darcy's law. the fluid discharge Q is given by:

$$Q = KIA \qquad (3)$$

and from Ohm's law:

$$J = \sigma E \qquad (4)$$

Where K is the hydraulic conductivity, I is the hydraulic gradient, A is the area of cross-section perpendicular to the direction of flow, J is the current density, E the electrical field intensity and σ the electrical conductivity ($\sigma = 1/\rho$, where ρ is the resistivity). Taking into account a prism of aquifer material having unit cross-sectional area and thickness h, Niswas and Singhal (1981) combined equations (3) and (4) to get

$$Tr = K\sigma T = \frac{KS}{\sigma} \qquad (5)$$

Fig. 5

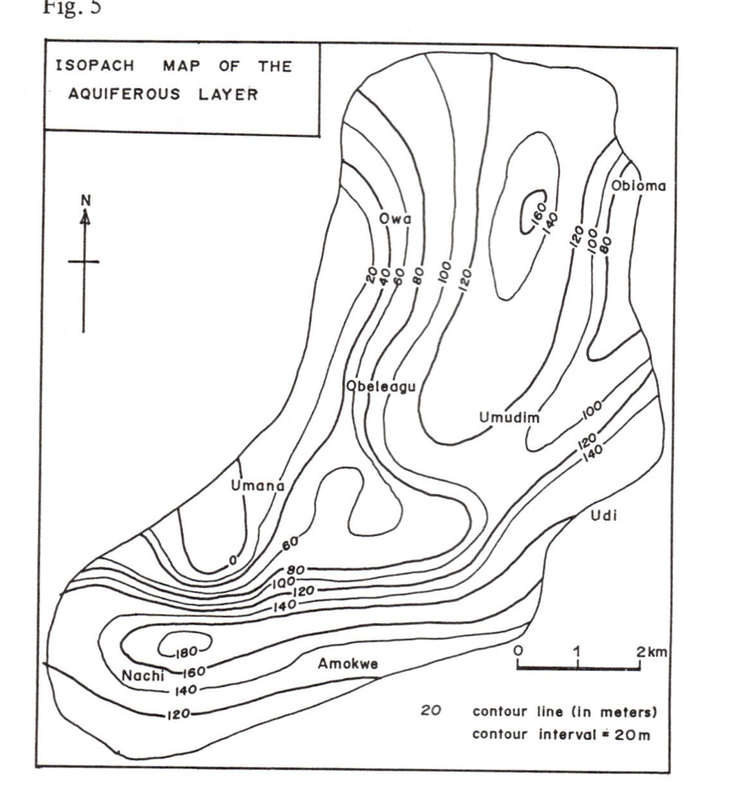
ISOPACH MAP OF THE AQUIFEROUS LAYER
N
Owa
Obioma
Obeleagu
Umudim
Umana
Udi
Nachi
Amokwe
20
40
60
80
100
120
160
140
120
100
80
100
120
140
0
60
80
100
120
140
180
160
140
120
0 1 2km
20 contour line (in meters)
contour interval = 20m

Fig. 6

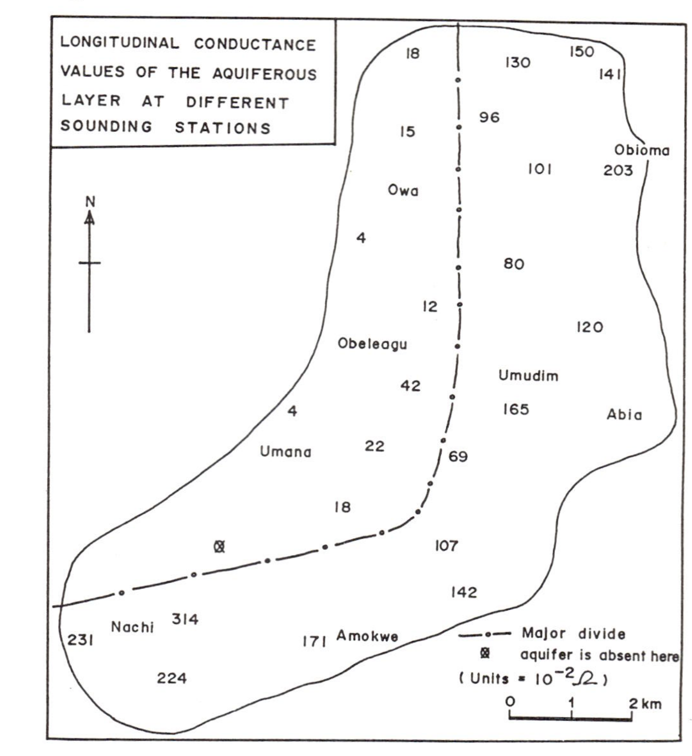
LONGITUDINAL CONDUCTANCE VALUES OF THE AQUIFEROUS LAYER AT DIFFERENT SOUNDING STATIONS
N
18
130
150
141
15
96
Obioma
101
203
Owa
4
80
12
120
Obeleagu
Umudim
42
165
Abia
4
Umana
22
69
18
107
142
231
Nachi
314
171
Amokwe
224
Major divide
aquifer is absent here
(Units = $10^{-2}\Omega$)
0 1 2km

Where Tr is the aquifer transmissivity (obtained by multiplying h with K). T is the aquifer transverse resistance, while S is its longitudinal conductance.

Figure 6 shows that around Nachi, Amokwe and Umudim the aquifer S values are high (1.07 - 3.14 Ω^{-1}), whereas they are low (0.04 - 0.22 Ω^{-1}) around Umana and Imezi Owa. The eastern half of the study area and other zones of high S values are probably underlain by very thick layers of conducting sediments. These areas (Nachi-Amokwe-Udi-Obioma) to the southern and eastern part of the study area have the best prospects for groundwater production - a fact already evident from Figure 5. Figure 6 thus shows that the area can be divided into two distinct zones on the basis of the aquifer longitudinal conductance values. The dividing line which cuts the area into an eastern and a western zone probably correlates with the position in the study area of the main watershed between the Anambra and the Cross River dranage systems to the west and east respectively.

It has been shown (e.g. by Niswas and Singhal, 1981) that in areas of similar geologic setting and water quality the product $K\sigma$ (hydraulic conductivity multiplied by the electrical conductivity) remains fairly constant. Knowing K values for the existing boreholes, and with σ values extracted from the sounding interpretation for the aquifer at borehole locations, it has been possible to determine transmissivity and its variation from place to place, including those places where no boreholes are available. Figure 7 shows the variation of transmissivities Tr for the aquiferous layer across the project area. Tr values are highest towards the eastern and south-eastern parts of the area.

On the basis of the $K\sigma$ product whose variation across the area is shown (Fig. 8), it can be concluded that the Amokwe-Nachi area (zone A) to the south is homogenous hydrologically, with $K\sigma$ values varying between 0.052 and 0.075 with a mean value of 0.066 and a standard deviation of 0.013. Udi, Obeleagu, Obioma and the Ninth Mile Corner areas appear to be homogenous also, but distinct from the other group with $K\sigma$ values ranging from 0.015 to 0.030 (with a mean value of 0.020 and a standard deviation of 0.004). Sufficiently high transmissivities coupled with good aquifer thicknesses for water well exploitation have made the vicinity of Nachi, Amokwe, Udi and the area to the west of Obioma the most prospective areas for the drilling of productive boreholes in the study area.

Table 1 summarizes the average values of some aquifer characteristics for the area. Data from the water well field at the Ninth Mile Corner to the north have also been included. From this table, it is obvious that the transmissivity values obtained from the pumping test data (parameter 2 on Table 1) represent those of the screened sections of the aquiferous layer in each borehole location. The true aquifer transmissivities (which are functions of the aquifer thickness) are given in the table as parameter 10. Thus, having obtained from the geophysical data, the variations across the study area of the thickness and transmissivity of the aquiferous layer (Figs. 5 and 7), it has been possible to deduce the average values of the hydraulic conductivities

Fig. 7

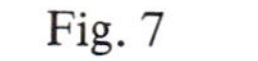

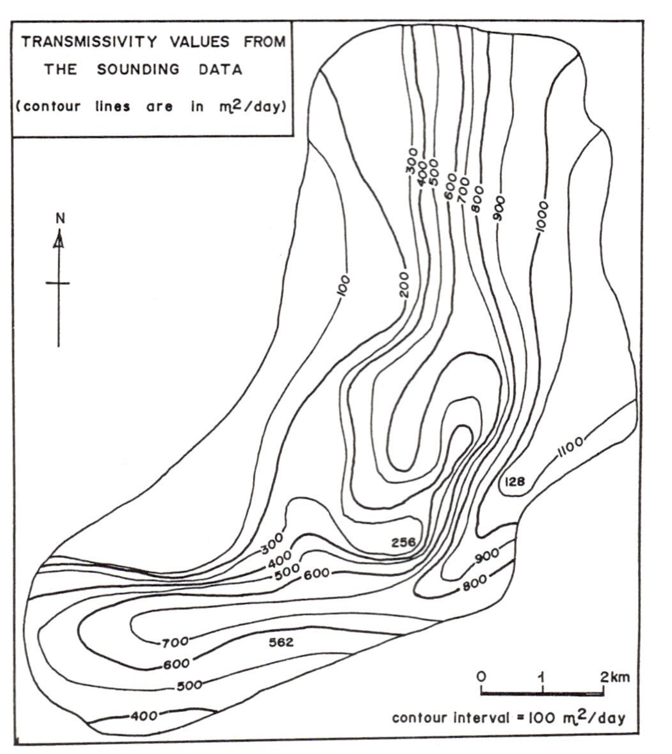
TRANSMISSIVITY VALUES FROM THE SOUNDING DATA
(contour lines are in m2/day)
N
100
200
300
400
500
600
700
800
900
1000
1100
128
256
300
400
500
600
900
800
700
562
600
500
400
0 1 2km
contour interval = 100 m2/day

Fig. 8

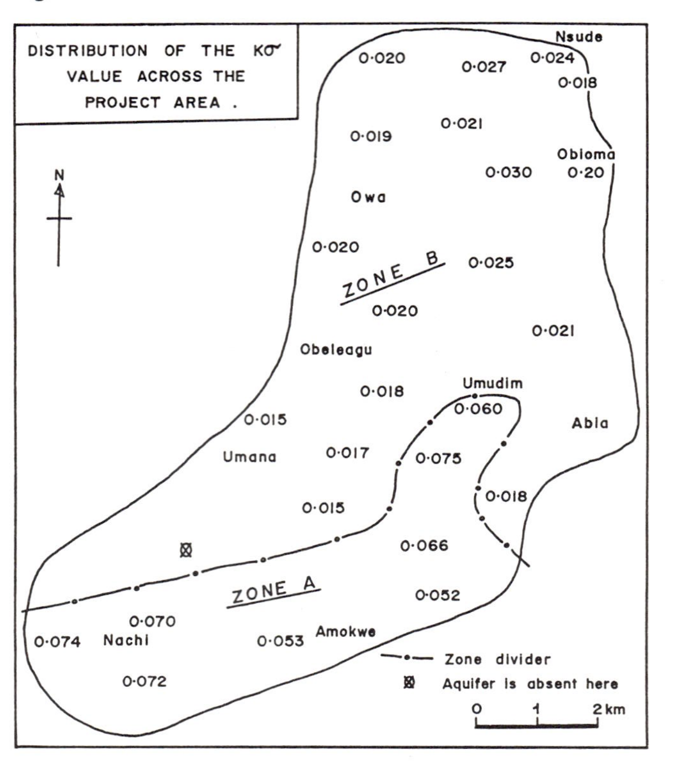
DISTRIBUTION OF THE Kσ VALUE ACROSS THE PROJECT AREA .
N
Nsude
0·020
0·027
0·024
0·018
0·019
0·021
Obioma
0·030
0·20
Owa
0·020
ZONE B
0·025
0·020
0·021
Obeleagu
Umudim
0·018
0·060
0·015
Abia
Umana
0·017
0·075
0·018
0·015
0·066
ZONE A
0·052
0·070
0·074
Nachi
0·053
Amokwe
Zone divider
Aquifer is absent here
0·072
0 1 2km

Table 1: Aquifer Parameters for Some Locations in the Project Area

S/No.	Parameter	Site Locations				
		Nachi	Amokwe	Udi	Obeleagu	Ninth Mile Corner
1.	Screen Length (m)	30.00	12.60	16.15	16.80	33.48
2.	Transmissivity (m^2/day)	133.44	47.63	191.54	76.26	183.12
3.	Average hydraulic Conductivity (m/day)	4.45	3.78	11.86	4.54	5.47
4.	Thickness of aquiferous zone (m)	134.00	64.00	87.00	9.08	114.50
5.	Aquifer resistivity (Ω m)	60.00	60.00	705.90	256.00	200.00
6.	Transverse Resistance of aquifer (Ωm^2)	8040.00	3840.00	61413.30	2324.48	22900.00
7.	Aquifer longitudinal conductance (Ω^{-1})	2.23	1.07	0.12	0.04	0.57
8.	$K\sigma$ value	0.07	0.06	0.02	0.02	0.03
9.	K/σ value	265.87	254.37	8349.27	1172.24	1094.40
10.	Aquifer transmissivity (m^2/day)	577.85	251.28	1115.09	46.69	655.40
11.	Hydraulic conductivity (m/day)	4.31	3.93	12.82	5.14	5.72

Note: Of the 11 parameters, 1-3 represent basic data from the boreholes at the respective locations, while 4-11 were obtained from the field data. There is a close agreement between the theoretically deduced and the observed values of hydraulic conductivities at the borehole locations.

for each measurement station. These K values are not shown on a separate sketch, but parameter 11 on Table 1 shows the calculated values at the borehole locations. Actual values for K are available for these locations (from grain size analyses and pumping tests) and are given by parameter 3 on Table 1. The close agreement between these values is a good indicator of the reliability of the results presented here.

DISCUSSION AND CONCLUSION

Despite problems inherent in the conventional curve-matching techniques, the close agreement of the interpretation of the sounding data with geological information from the available boreholes gives an indication of the usefulness of the present project in helping to broaden our understanding of the hydrogeophysical properties of the aquifer systems in this area. Computer-aided direct interpretation methods would no doubt help in better resolving the thicknesses and resistivities of the subsurface layers but would not appreciably change the general conclusions and results given here. Drilling results from boreholes at the Ninth Mile and at Udi have revealed the presence of a multi-storey aquifer system in the area. In most cases the upper aquifer is separated from the lower one by thin bands of impermeable clay and clayey sands. These 2 - 5 m thick bands have not been "sensed" at all on the sounding curves because of the principle of suppression. This principle, which is an ambiguity that developes in the interpretation of electrical soundings, states that a layer whose resistivity is intermediate between those of enclosing layers is not visible on the electrical sounding unless it is of great thickness (Kunetz, 1966; Salát, 1968).

The present survey has helped to map out zones for the drilling of productive boreholes in the study area. Potential aquifers occur at depths ranging between 12 and 195 m. Aquifer thickness is highly variable in the area, but appears thickest around Nachi, Amokwe and to the west of Obioma. Variations in hydraulic properties occassioned by changes in subsurface geology or water quality or both, occur. The Nachi-Amokwe area appears homogenous in this respect and distinct from the Obeleagu - Obioma - Ninth Mile areas which again appear homogenous, but different from the earlier zone. The diagnostic features of the $K\sigma$ product which have earlier been discussed in detail by Niswas and Singhal (1981) have also proved very useful in this study. The dearth of hydrochemical information on waters from the whole area makes it difficult to confirm at present any water quality variations that maybe present.

ACKNOWLEDGEMENT

The authors are grateful to K.O. Uma of the Geology Department, UNN for supplying pumping test data from boreholes in the study area, and for very useful discussions.

REFERENCES

Aka, E.U. and K.M. Onuoha, 1983: Groundwater investigation in the Enugu area by the electrical resistivity method. Paper presented at the 19th Annual Conf., N.M.G.S., Warri.

Hazell, J.R.T., 1955: The Enugu ironstone, Udi Division, Onitsha Province, Rec. Geol. Surv. Nigeria, 44-58.

Henriet, J.P., 1976: Direct applications of the Dar Zarrouk parameters in groundwater surveys. Geophys. Prospecting, v. 24, 344-353.

Kunetz, G., 1966: Principles of Direct Current Resistivity Prospecting. Gebrüder Borntraeger, Berlin-Nikolassee.

Mbazi, F.C.C., 1985: Electrical Resistivity Survey for Groundwater in Parts of Udi and Ezeagu L.G.A's of Anambra State, Nigeria. Unpublished M.Sc. thesis, Univ. of Nigeria, Nsukka.

Niswas, S., and D.C. Singhal, 1981: Estimation of aquifer transmissivity from Dar Zarrouk parameters in porous media. J. Hydrology, v. 50, 393-399.

Salát, P., 1968: Electrical Methods of Geophysical Prospecting. ELTE University Textbook, Tankonyvkiado, Budapest (in Hungarian).

Zhody, A.A.R., 1965: The auxiliary point method of electrical sounding interpretation and its application to the Dar Zarrouk parameters. Geophysics, v. 30, 644-660.

Zhody, A.A.R., 1974: Electrical resistivity method in groundwater investigations. In: Applications of Surface Geophysics to Groundwater Investigation. Book 2, Chap. 1., United States Geol. Survey.

Hydrogeological Deductions from Four Geoelectrical Profiles in Calabar, Nigeria

C. I. Adighije[1] and P. O. Okeke[2]
[1]Cinab Engineering and Geological Services Ltd., P.O. Box 1277, Owerri, Nigeria
[2]School of Natural and Applied Sciences, Federal University of Technology, P.M.B. 1526, Owerri, Nigeria

Keywords: Geoelectric data, Stratigraphic interpretation, Electrical conductivity, Multi-layered aquifers

ABSTRACT

Geoelectric data and stratigraphic sub-surface cross-section from the Pleistocene Coastal Plain Sands of Calabar north, Nigeria are presented. The results from twelve VES points, using Schlumberger confriguration indicate three major sub-surface rock units:

Clayed cover (<10 m thick) underlain by fine clayed sands (>10 m - 50 m thick; 2,600 - 7000 ohm-m), and coarse dands, the aquifer zone (>100 m thick; 200 - 700 ohm-m) in that order. A major hydrogeological deduction from this stratigraphic approach in the interpretation of resistivity data, is that productive water boreholes are feasible within the study area. Recommended yields of the boreholes range from medium to low (average: 2 - 5 litres/s).

INTRODUCTION

There are many applications of geophysics in water supply and control problems. Control includes the design and construction of irrigation and drainage projects, such as canals and aqueducts, and in the planning of dam and reservoir sites for flood control and power projects - engineering problems (Dunning, 1970). Prospecting for water is essentially a geological problem, and the geophysical approach is dependent on the mode of the geological occurrence of water (Bhattacharya and Patra, 1968). Water may occur as ground water proper, fissure (Clark, 1983), mineral springs, cavern water and water issuing out of the leaks.

Geophysically, the location of groundwater may be determined in 3 ways - direct, stratigraphic and structural. Direct methods are largely confined to well-logging, or locating the sites of radioactive and thermal springs (Scott, Keys and MacCary, 1971). The stratigraphic method implies locating water-bearing formations through distinguishing physical properties imparted by the presence of water, such as high seismic velocity, or increased or decreased electrical conductivity. The structural approach means the mapping of key beds that bear a certain stratigraphic or structural relating to the water-bearing bed ie. mapping of synclines, erosional troughs or structural lows.

In this paper, the stratigraphic method of location is adopted because the water-bearing strata should differ in electrical conductivity,atleast, from the other formations in a rock series. Although seismic refraction can be used for locating the ground-water table (because of the differences in elasticity for different media) (Eatan and Watkins, 1967), the Vertical Electrical Sounding (VES) is preferred and selected for this study for two reasons:

1) Traditionally, with the Schlumberger configuration (preferably), it is the most suitable method for ground water investigation in most geological occurences, and
2) It is comparitively low in cost of exploration.

This study preceeded the water borehole drilling project at the Cross River Basin Development Authority headquarters (Ikot Effanga, Calabar).

GEOLOGY AND HYDROGEOLOGY

The study area (Calabar; Fig. 1) is within the Pleistocene to Pliocene Coastal Plain Sands (and Clays) overlying Detritic sediments of the Tertiary Ameki Formation. The Ameki Formation (lateral equivalent of Benin Formation) is a heterogeneous sequence of sediments, ranging from the fine to course sands, intercalated with calcareous shales and clays, and with a generally high clayey component.

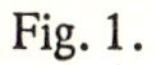

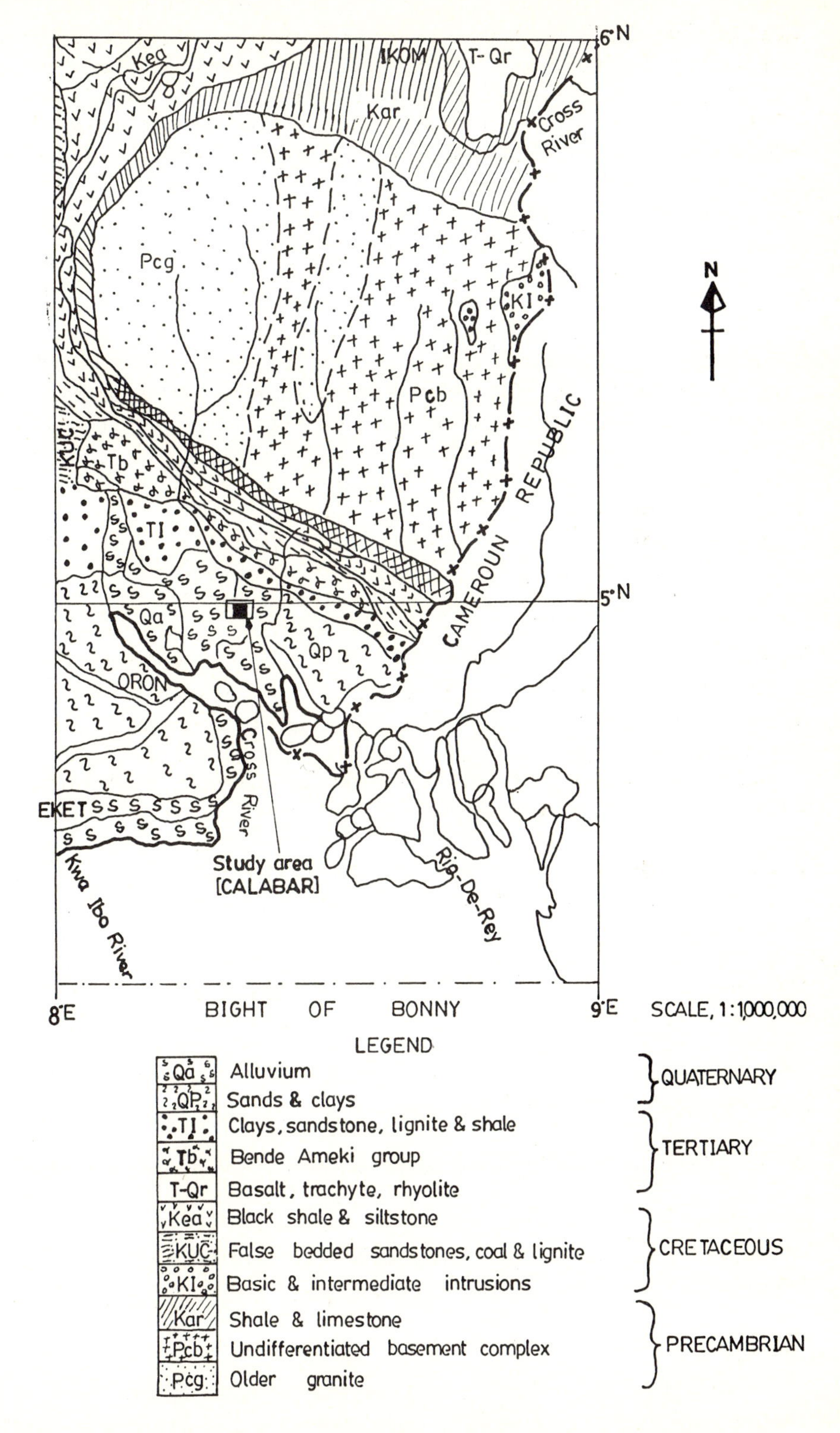

6°N
IKOM
T-Qr
Kea
Kar
Cross River
Pcg
Pcb
KI
KUC
Tb
Tl
REPUBLIC
CAMEROUN
5°N
Qa
Qp
ORON
Cross River
EKET
Study area
[CALABAR]
Kwa Ibo River
Rio-De-Rey
N
8°E
BIGHT OF BONNY
9°E
SCALE, 1:1,000,000
LEGEND
Qa Alluvium
QP Sands & clays
QUATERNARY
Tl Clays, sandstone, lignite & shale
Tb Bende Ameki group
T-Qr Basalt, trachyte, rhyolite
TERTIARY
Kea Black shale & siltstone
KUC False bedded sandstones, coal & lignite
KI Basic & intermediate intrusions
CRETACEOUS
Kar Shale & limestone
Pcb Undifferentiated basement complex
Pcg Older granite
PRECAMBRIAN

The two geological formations are important as extended multi-layer aquifers, of medium to low yields. Because of the high clayey component and heterogeneous nature of sand in the study region, transmissivity of the aquifers is low when compared with the Benin Formation. These low transmissive multi-layered aquifers are easily blocked by thick mud-cakes due to excessively dense drilling muds and prolonged drilling ratios. Development of boreholes could, therefore, be difficult in such geological environments.

THE INVESTIGATION

Direct current VES with Schlumberger configuration was adopted in the investigation. ABEM DC Type 5310 Terrameter was used. Minimum depth of probe was 100 metres. Twelve soundings were distributed in four geoelectric profiles within the study area. Details of the VES separation are given below:

Profile I

VES/1 to VES/2 = 50 m separation
VES/2 to VES/3 = 50 m separation
VES/3 to VES/4 = 150 m separation

Profile II

VES/8 to VES/9 = 50 m separation
VES/9 to VES/10 = 75 m separation

Profile III

VES/5, VES/6 and VES 7 were separated by 50 m

Profile IV

VES/11 and VES/12 were separated by 30 m

RESULTS

Numerical results from the twelve VES points were obtained by Ethel Bert Auxilliary point method using the appropriate matching curves. These are presented in Table I, and are used in constructing resistivity cross-sections (Figs. 2 and 3).

Figures 2 and 3 reveal three major rock horizons:

1) Clayey cover (< 10 m thick),
2) Fine clayey sands (> 10 m - 50 m thick; 2,600 - 7000 ohm-m), and
3) Coarse sands (> 100 m thick; 200 - 700 ohm-m).

The Coarse sands form the aquifer zone.

Table 1 Geoelectric Data

VES	Depth(m)	Resistivity (Ω-m)	VES	Depth(m)	Resistivity (Ω-m)
1	0 - 3.6	1200	7	0 - 2.6	700
	3.6 - 40	3200		2.6 - 68	5000
	40	550		68	600
2	0 - 3.4	750	8	0 - 4.5	700
	3.4 - 55	4000		4.5 - 13.5	7000
	55	550		13.5	200
3	0 - 4.8	1600	9	0 - 3.0	650
	4.8 - 53	4500		3.0 - 18	6000
	53	700		18	200
4	0 - 4.2	1000	10	0 - 1.2	550
	4.2 - 46	2600		1.2 - 18	5600
	46	400		18	200
5	0 - 3.0	400	11	0 - 7.0	1000
	3.0 - 75	4000		7.0 - 16	5000
	75	400		16	580
6	0 - 2.8	580	12	0 - 7.5	1300
	2.8 - 78	3500		7.5 - 13	7000
	78	650		13	480

Fig. 2

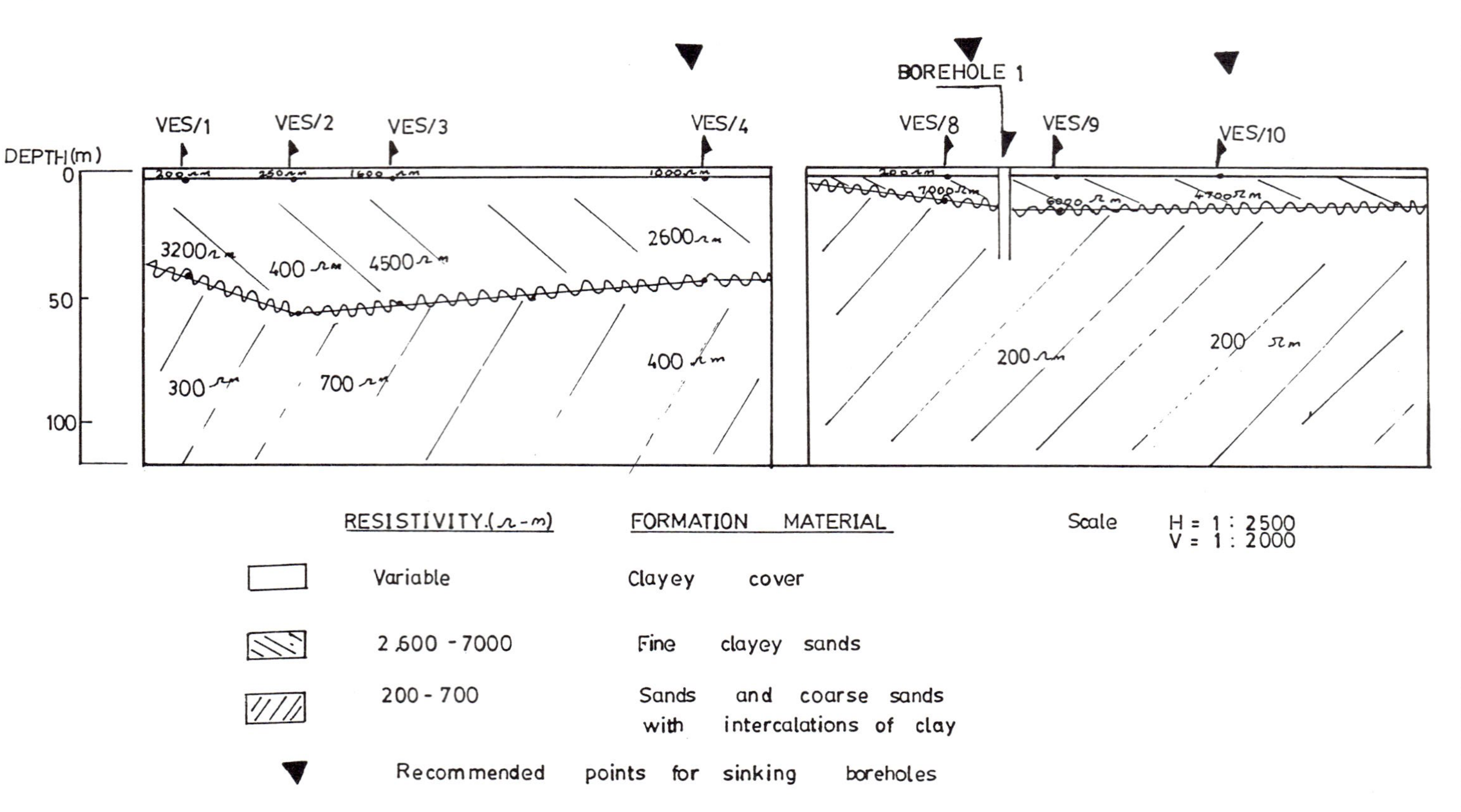

DEPTH (m)
0
50
100
VES/1
VES/2
VES/3
VES/4
VES/8
VES/9
VES/10
BOREHOLE 1
3200 Ω m
400 Ω m
4500 Ω m
2600 Ω m
300 Ω m
700 Ω m
400 Ω m
7000 Ω m
6000 Ω m
4700 Ω m
200 Ω m
200 Ω m
RESISTIVITY (Ω - m)
FORMATION MATERIAL
Scale
H = 1 : 2500
V = 1 : 2000
Variable
Clayey cover
2,600 - 7000
Fine clayey sands
200 - 700
Sands and coarse sands with intercalations of clay
Recommended points for sinking boreholes

Fig. 3

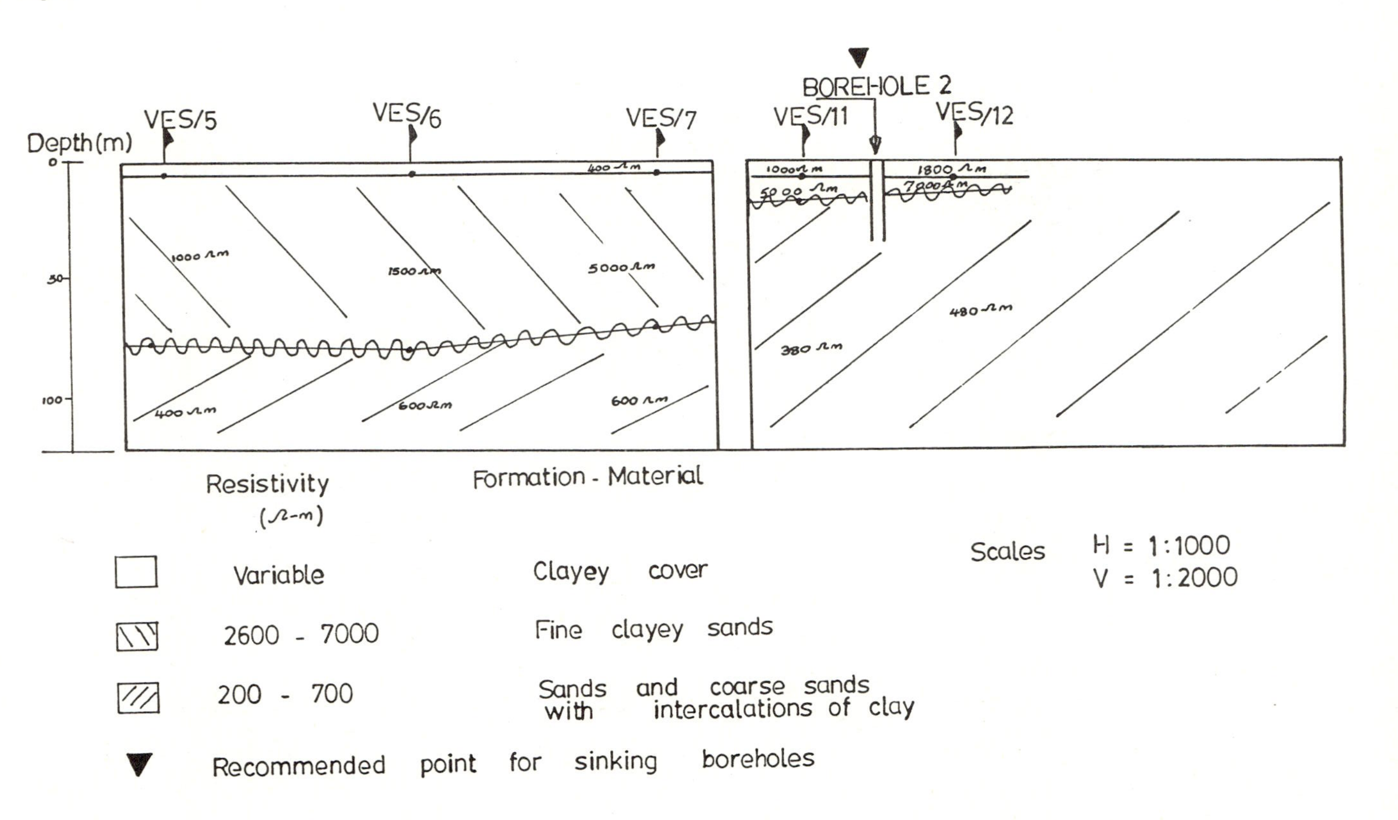
BOREHOLE 2
VES/5
VES/6
VES/7
VES/11
VES/12
Depth(m)
0
50
100
400 Ωm
1000 Ωm
1500 Ωm
5000 Ωm
400 Ωm
600 Ωm
600 Ωm
1000 Ωm
5000 Ωm
1800 Ωm
7000 Ωm
380 Ωm
480 Ωm
Resistivity (Ω-m)
Formation - Material
Variable
Clayey cover
2600 - 7000
Fine clayey sands
200 - 700
Sands and coarse sands with intercalations of clay
Recommended point for sinking boreholes
Scales
H = 1:1000
V = 1:2000

HYDROGEOLOGICAL DEDUCTION:

Considering the available geological and geophysical data, the following hydrogeological recommendations can be made with confidence:

1) All th VES points are favourable for the location of productive borehole, since the coarse sands (aquifer) are common in the entire area.

2) Pumping test data, available with the first author, indicate that expected yields of boreholes within the study area range from medium to low (average; 2-5 litres/s).

3) Boreholes should be drilled according to the following specifications:
Drilling system: Rotary (Direct or Reverse circulation).
Casing Diameter: Minimum 8" Depth of Borehole: 300 ft (100 m).
Filters: 18 m sand-screen (3 lengths).
Pre-Filters: Gravel-pack with calibrated gravels.
Developing of well: Minimum 24 hrs. continuous extraction.

ACKNOWLEDGEMENT

We thank CINAB Engineering and Geological Services for the permission to publish the data presented in this paper, and for the use of their geophysical and geological facilities.

REFERENCES

Clark, R.D., 1983: Geophysical Exploration = No Water Shortage? Geophysics: The Leading Edge of Exploration. July, 30-34.

Dunning, K.C., 1970: Geophysical Exploration. An outline of the principal method used in the exploration for minerals, oil, gas and water supplies. Publication of the Institute of Geological Sciences, The Geological Museum, London.

Eaton, G.P. and J. Watkins, 1967: The use of seismic refraction and gravity methods in hydrogeological investigations. In: Mining and Groundwater Geophysics. Geological Survey of Canada Economic Report No. 26.

Scott Keys, W. and L.M. Mac Cary, 1971: Application of Borehole Geophysics to Water-Resources investigations. Chapter E1, Book 2. Techniques of Water - Resources Investigations of the United States Geological Survey.

Groundwater Fluxes and Gully Development in S. E. Nigeria

Kalu O. Uma and K. M. Onuoha
Department of Geology, University of Nigeria, Nsukka, Nigeria

Keywords: Gullying, Erosion, Mass Wasting, Groundwater, Hydraulic gradient, Landsliding, Seepage force

ABSTRACT

Active gullying involving accelerated erosion, bank erosion and mass wasting, has caused tremendous destruction of life and property in many areas in southeastern Nigeria. The major mechanism of gullying in the most dangerous spots is mass wasting involving landslides, and soil creep. These are caused by high groundwater levels in the gully slopes and the consequent high seepage forces and pore water pressures. The values of the seepage forces at the active gully slopes commonly increase to about 10 to 100 times their normal values at the stable areas. These and other hydrogeotechnical factors acting in combination reduce the factor of saftey of the gully slopes below critical levels and cause the eventual and frequent failures of the slopes.

Any control measure against the onslaught of the gullying must encourage the permanet dewatering of the gully slopes and lowering of the groundwater levels in the vicinty of the gullies. Measures that tend to force more infiltration of rainwater, such as those currently being applied in the area should be stopped.

INTRODUCTION

Gullies are widespread in Nigeria especially in the southeastern region. However, the intensity of gullying and the erosion phenomena associated with it vary from place to place. Based on the predominant erosion mechanism, the active gully spots can be classified into 3 major groups. The first group is composed of those gullies in which the dominant process is accelerated erosion: this is the classical gully erosion - sensu stricto. In the second group, bank erosion with the associated undercutting is the main phenomenon, while the third group of gully spots comprises those places where mass wasting (landslides, slumping and soil creep) is dominant. Gullies of the first group are often shallow and surfical and are seldom more than 10 m deep. They may or may not be associated with streams or rivers. The second group of gullies are found at the banks of mature or semi-mature rivers and are frequently due to undercutting at the toe of the banks by flood water.Gullies in the third group often start off as those in group one but subsequently cut deep into the bedrock (below the water table) and have groundwater seepages at various levels of their banks. They are normally deep (> 20 m) and processes going on in them involve sudden movements of large masses of earth materials through landslides and slumping. Gullies of the third group constitute the most serious threat to life and property.

This paper is concerned, primarily, with gullies of the third group (where mass wasting is the major process). These gullies are often associated with those of groups one and two and the major erosive processes acting in combination have wrecked homes, swept crops and washed roads away. Figure 1 shows the distribution of active gully areas in southeastern Nigeria. Gullies of the third group are found in (1) the banks of the Idemili and Orashi Rivers and their headstreams; (2) the Agulu-Nanka-Oko area: (3) the Ukehe-Udi area; and (4) the Abiriba-Ohafia area.

Assessment of Previous Studies and the Measures Taken to Arrest Gully Development

The Agulu-Nanka-Oko gully complex is easily the most threatening environmental hazard in the study area and has been subjected to a series of qualitative and semiquantitative studies by government agencies and institutional researchers. From the much that has been postulated by different workers on the origin and development of the gullies, there appears to be a considerable measure of agreement. Table 1 summarises the opinions of the main workers. They all (except Egboka and Nwankwor, 1985) seem to emphasise the importance of the soil and geologic materials exposed by the removal of vegetation cover and the impact of heavy rainfall on such materials. The concensus of these earlier workers is that the high intensity rainfall in the area produces high volumes of overland flow with high erosive energy. The action of the high erosive floods on the unusually highly susceptable geologic and soil materials produce the complex gullies. Floyd (1965) went further to suggest a 6-stage evolution of the gullies. These are:

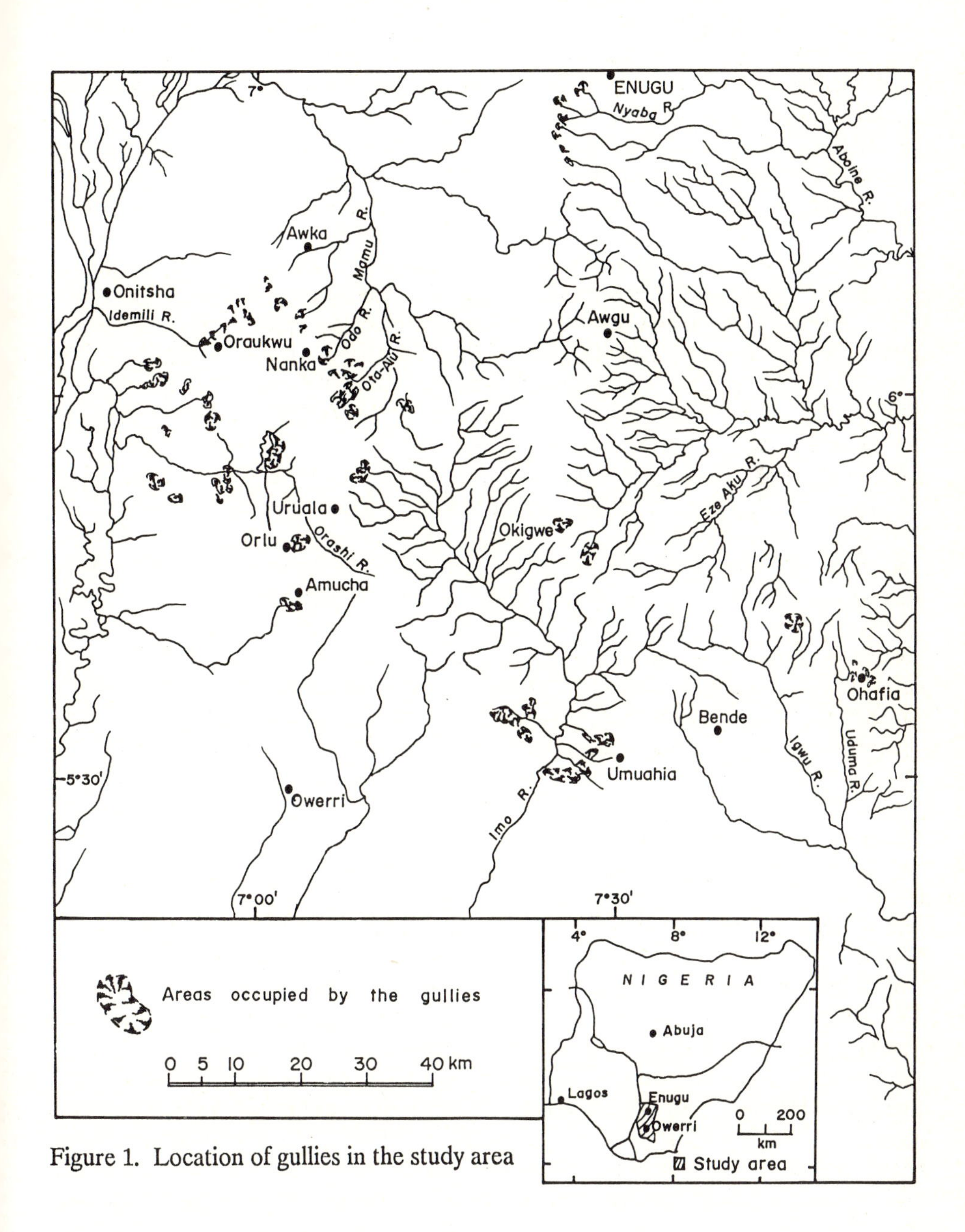

Figure 1. Location of gullies in the study area

Table I: Summary of Suggested Origin of Gullies in Southeastern Nigeria

Author	Gully System Studied	Suggested Cause(s)	Other Remarks
Floyd (1965)	Agulu-Nanka-Oko and Ukehe-Udi	Anthropic factors; removal of vegetation cover and exposure of the feraltic sandy soils to direct impact of rain.	Mechanism of erosion is classical gullying (accelerated erosion) and mass wasting (sliding and slumping).
Ofomata (1965)	Enugu area (Agulu-Nanka-Oko and Enugu Udi area)	Combined effect of concentrated floodwater and low resistive soils and geologic materials	Geologic and soil materials are structurally stable. Thus splash erosion is not a major factor.
Ogbukagu (1976)	Agulu-Nanka-Oko-and Alor-Oraukwu area.	Concentrated runoff and groundwater seepage. High susceptibility to erosion due to loose nature of materials.	
Technosynesis (1978)	All gullies in Anambra and Imo states.	Overlandflow (run-off)	Major processes are classical gully erosion bank erosion and landslides/slumping
Nwajide and Hoque (1979)	Agulu-Nanka-Oko	Physical factors; high erodibility of the geologic and soil materials and the topography of the area. Biotic factors; Burrowing activities of animals and plant roots Anthropic factors; Farming, Civil works and deforestation	Major processes are rain splash, pluviofluvial scouring, groundwater flow and mass wasting.
Egboka and Nwankwor(1982)	Agulu-Nanka-Oko	Geotechnical factors; pore water pressure and Hydrogeochemistry of the groundwater environment	There is loss of shear strength due to increases in pore pressure. This leads to piping, cave-ins, sliding and slumping.

Stage 1: Intensive agriculture leading to soil degradation and destruction

Stage 2: Rain splash and removal of soil particles

Stage 3: Leaching and eluviation

Stage 4: Sheet and rill erosion

Stage 5: Accelerated erosion and formation of gullies

Stage 6: Mass earth movement through slumping, sliding and downhill creep leading to the complex badlands.

Some of the gullies evidently have not undergone the complete evolution sequence and some have stabilized before the last stage.

The earlier studies seem to have concentrated on the mechanisms of classical gullying and bank erosion. Thus the factors that accelerate classical gullying and bank erosion have been meticulously analysed and highlighted. The palliative measures recommended also rely heavily on the assumption that overland flow is the major agent. The measures tended to suggest a reduction of the quantities of floodwater flowing in the drainage network or the reduction of their veocities. Based on the findings and suggestions of these workers, a network of check dams, embankment and consolidation works were designed by the Technosynesis, planning and engineering consultants working for the Federal Ministry of Agriculture and Rural Development in 1978. Intensive reafforestation was also recommended. These measures apart from reducing the volume of floodwater and/or their velocities were also intended to force more rainwater into the subsurface.

In the Agulu-Nanka-Oko, Alor-Oraukwu and Amucha areas, diversion channels have been built to divert floodwater from the major gully heads. The result of these measures has been very precarious and of no lasting duration. In some gullies around Oraukwu, Awgbu and Alor the reduction in flood volume/velocity has apparently stabilized the gullies but in most of the dangerous spots around Agulu, Nanka, Oko, and Amucha, sliding and slumping have continued unabated. Earlier control measures involving check dams, infilfration pits, reafforestation etc., also failed to stop the development of the gullies. Some structural framework of the check dams are still preserved at the bottom of the gully in the slided chunk of earth materials. Although many excuses have been given for the failure of these control measures, the undeniable implication is that the measures were not enough and did not consider all the elements responsible for the gullying. It has become necessary therefore, to search for and quantitatively analyse the missing elements.

In the case of the third group of gullies, the dominant phenomena involve sliding and slumping and could thus be treated as a slope problem. Unfortunately the earlier workers have tended to ignore this; consequently, control measures based on their recommendations have failed and probably would continue to fail. The gullies can only be checked when all the aspects of the erosion are considered in the design of the measures.

It is generally accepted by engineers and environmental scientists that slope stability analyses require an understanding of pore water pressures and seepage forces in a slope. The understanding must be based on a knowledge of the groundwater flow systems which in turn depend on the regional geologic environment. We will now turn our attention to the effects of the groundwater fluxes on the stability of the gully slopes and thus on the development of the gullies. An attempt would be made to highlight the pattern of groundwater fluxes in the area and to relate them to the pore pressure build up and seepage forces acting in the geologic materials. The effects of these hydrogeotechnical factors on the activating forces generating the gullies would also be investigated. It is our conviction, that a better understanding of the dynamics of gullying in the area cannot be achieved without taking these factors into consideration. In designing lasting control measures, their effects are critical.

PHYSIOGRAPHIC AND CLIMATIC FACTORS

Generally speaking the active gullies do not coincide with a particular topographic form, although most are associated with cuestas. The Agulu-Nanka-Oko gully complex is found on the east-facing scarp slope of a roughly N-S trending cuesta. The crest of the cuesta stands at over 350 m above mean sea level (Nwajide, 1977). The dip slope is towards the west-southwest, merging imperceptibly with the River Niger valley. The Ukehe-Udi gullies developed on the crest of the Enugu-Udi cuesta. The crest of this escarpment forms the watershed between the Niger and Cross River drainage systems. Most of the stream channels on the dip slope are deeply incised, and intense gullying of very recent origin has taken place at many points mainly in the outcropping loose sandstones. In the Abiriba-Ohafia area, the gullies are also found on the dip slopes of the southern part of the Enugu-Udi escarpment. However, the active gullies occur both on the scarp and dip slopes of the escarpment. The other gullies developed on more or less gentle slopes but have cut deeply enough into the bedrock to form significant topographic features. The gullies generally have a depth range of 5 m to 50 m and carry ephemeral and sometimes perennial springs fed by groundwater and sporadic surface runoff from the adjacent uplands.

The active gully areas fall within the rain forest and savannah belts of Nigeria. The rainfall regime is characterised by a dry season from November to March and a rainy season between April and October. The total annual rainfall incrases continuously from the interior towards the coast. In the study area it varies from 1800 mm at

Enugu, in the northern section, to over 2300 mm at Owerri, at the southern boundary. The number of rainy days also increases in the same direction. The characteristic intensity - duration relationship of the rainfalls typical of the area is shown on Table 2. The table also shows the erosivity of the rains based on empirically derived kinetic energy index (KE > 25). The KE > 25 index gives the total kinetic energy of rains with mean intensities greater than 25 mm hr^{-1}; 25 mm hr^{-1} is the threshhold value below which rains (in tropical African environments) can be considered as non-erosive (Technosynesis, 1978). Using the KE > 25 index, the kinetic energy developed by single storms in the area varies from 89.0 to over 1,950 joules m^{-2}. Such heavy rains falling on the bare land (exposed by cultivation) could tear up soils which are even more structurally stable than those in the gully-prone areas. This has no doubt aroused the interest of many researchers especially those with a geomorphological background.

GEOLOGY AND HYDROGEOLOGY

The active gully areas are underlain by two major geological formations. The Ukehe-Udi, and the Abiriba-Ohafia areas are underlain by the Ajali Sandstone of Maestrichtian age. The Ajali Sandstone consists of a great thickness of friable, poorly sorted sandstones, typically pure white in colour, but sometimes ironstained (Simpson, 1954). A marked banding of coarse and fine layers and large scale cross-bedding are the characteristic sedimentary structures of the formation. The angle of inclination of the foreset laminae with the underlying major bedding planes ranges up to 20°. Thin bands of white mudstone and shale occur at intervals increasing in number towards the base. The Ajali Sandstone is underlain and overlain by the Mamu and Nsukka Formations respectively. A considerable thickness of laterite formed by the weathering and ferruginization of the sandstone caps this formation. The lateritic red earth is sandy and contains clay fractions varying from 4 % on the surface to 30 % at about 4 m below the surface (Ofomata, 1965).

The Agulu-Oko and Alor-Oraukwu areas are underlain by the Nanka Sands of Eocene age. The Nanka Sand is a sequence of unconsolidated or poorly consolidated sands, 305 m thick. It is underlain by the Paleocene Imo Shale and overlain by the Ogwashi/Asaba Formation of Pliocene-Oligocene age. The beds are predominantly sandy with minor claystone and siltstone bands. The sands are poorly sorted, cross bedded and medium to coarse-grained.

Although the formations underlying the active gully areas differ, the materials show a high degree of similarity in lithology. Sandstone forms the dominant units which are subdivided by thin claystone units at various horizons. There is the virtual absence of cement and the consequent friability of the units. This accounts for their susceptibility to erosion. Grain size data of samples from the gullies (Ofomata, 1965; Technosynesis, 1978; Nwajide and Hoque, 1979) show a preponderance of medium to coarse (0.25 - 1.00 mm) grain size and paucity of fines and matrix. The clay and

Table 2 Characteristic Intensity-Duration Relationship of Rainfalls typical of the study area
(Modified from Technosynesis, 1978)

Day	Total Rainfall (mm)	Rainfall Intensity for the first 30 mins (mm)	I_{mean} cm/hr	Total Kinetic * Energy (K E) Joules/m²	KE>25 Joules/m²
1	1.02	1.02	0.584	19.32	-
1	0.64	0.64	0.730	12.67	-
2	10.80	8.64	1.188	234.20	-
3	4.70	4.70	1.801	109.48	-
4	79.00	33.02	2.583	195.03	195.03
	15.24	3.10	0.610	291.12	-
5	0.64	0.64	0.349	10.85	-
6	0.64	0.64	0.365	10.96	-
	0.64	0.64	0.365	10.96	-
	0.64	0.64	0.730	12.67	-
7	14.86	14.86	4.086	393.20	393.20
	1.27	1.27	0.254	19.9	-
12	39.37	34.13	3.937	1036.0	1036.1
	6.99	2.54	0.279	112.44	-
13	0.64	0.64	0.699	12.57	-
15	9.27	6.50	1.28	203.56	-
	2.54	1.30	0.254	39.94	-
16	4.70	4.70	3.759	122.85	122.85
17	1.90	1.00	1.190	27.74	-
20	2.54	2.60	0.389	44.12	-
22	5.46	5.46	3.003	137.98	137.98
27	3.43	3.43	0.985	71.90	-
	1.27	1.27	0.254	19.97	-
	10.80	10.80	4.138	286.30	286.30
28	0.64	0.64	0.699	12.57	-
	3.43	3.43	1.314	75.72	-
31	64.77	45.72	5.481	1787.38	1787.38
	1.27	0.26	0.051	12.01	-
	290.38			7095.65	5714.16

*total KE = (210.2 + 89 log I_{mean}) x total centimeter of rainfall.

I_{mean} is the mean rainfall intensity per hour

siltsizes (< 0.063 mm) are often less than 5 %. Both formations are capped by thick mantle of laterite formed by the leaching and ferruginization of the sandy units (Simpson, 1954; Nwajide, 1977). The loss of structure by lateritization has also contributed to further loss of strength and resistance to erosion. The soil/geologic framework thus provides a favourable environment for severe erosion.

The sequence of thick sand units separated by minor claystone/siltstone horizons form multi-aquifer systems. Akujieze and Ogbukagu (1985) indentified 4 major aquiferous horizons in the Agulu-Oko-Oraukwu area. The fourth (basal) aquifer rests on the Imo Shale. It occurs at depth ranges of 174 m at Alor and 270 m at Igboukwu-Oko area and is the most prolific aquifer in the area. However, the uppermost aquifer which is unconfined throughout the area is of high geotechnical interest. It is tapped by several springs and seepage zones. These springs and seepage faces are widespread on the gully banks at Agulu, Nanka, Oko and Oraukwu. The hydraulic properties of this aqufer have not been worked out, but grain size data from samples taken from the gully walls (along the seepage faces) indicate that the hydraulic conductivity is in the order of 1.0 x 10^{-2} cm/s. The hydrogeology of the other gully spots are little known and there is paucity of data needed to effectively assess them. However, the existence of shallow unconfined aquifers in the Enugu-Udi area is shown from the few borehole logs and springs.

Figure 2 shows the hydraulic potential head distribution in the shallow unconfined aquifer of the Agulu-Nanka-Oraukwu area. The figure is based on spring heads, boreholes and topography of the area. The closer isopotential lines indicate greater flux of groundwater. The areas of highest fluxes (closest isopotential lines) were found to coincide almost exactly with the active gully heads. Figure 3 displays the distribution of hydraulic gradients, while Figure 4 (a, b and c) shows flow nets along the sections shown in Figure 2. The closer nets around the effluent zones indicate greater fluxes of groundwater. The upward direction of the groundwater fluxes at the effluent zones agree with the widespread occurrence of boiling conditions observed in the field. A comparison of Figures 1, 3 and 4 reveals that the areas of high hydraulic gradient and upward flux coincide with the most unstable spots. Figures 5, 6 and 7 are similar diagrams for the Enugu-Udi area. In this area, we observe exactly the same trend as in the Agulu-Nanka-Oraukwu area. The complex pattern of groundwater fluxes in the Enugu-Udi area similarly agrees with the complex pattern of gullying in the area.

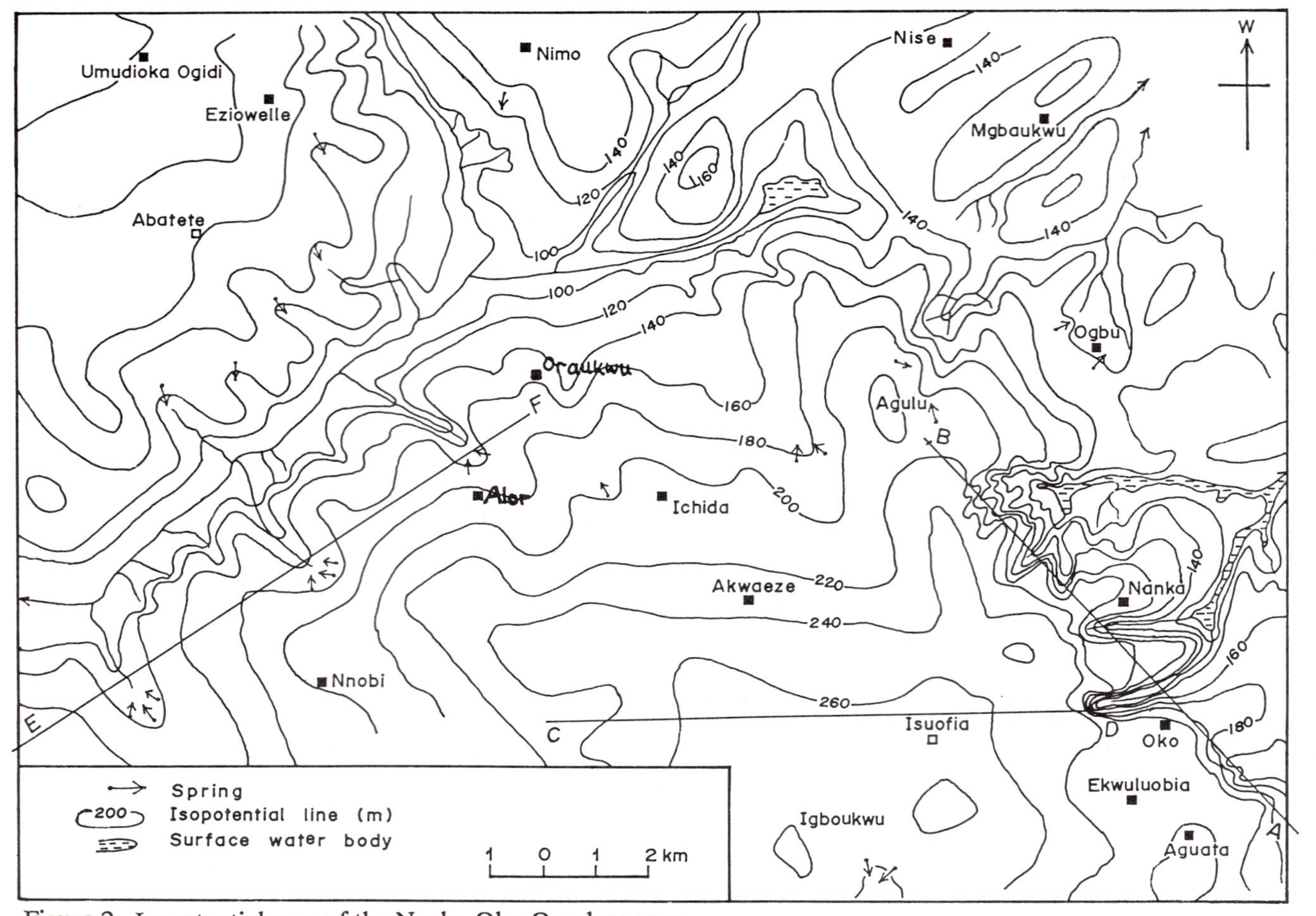

Figure 2. Isopotential map of the Nanka-Oko-Oraukwu area

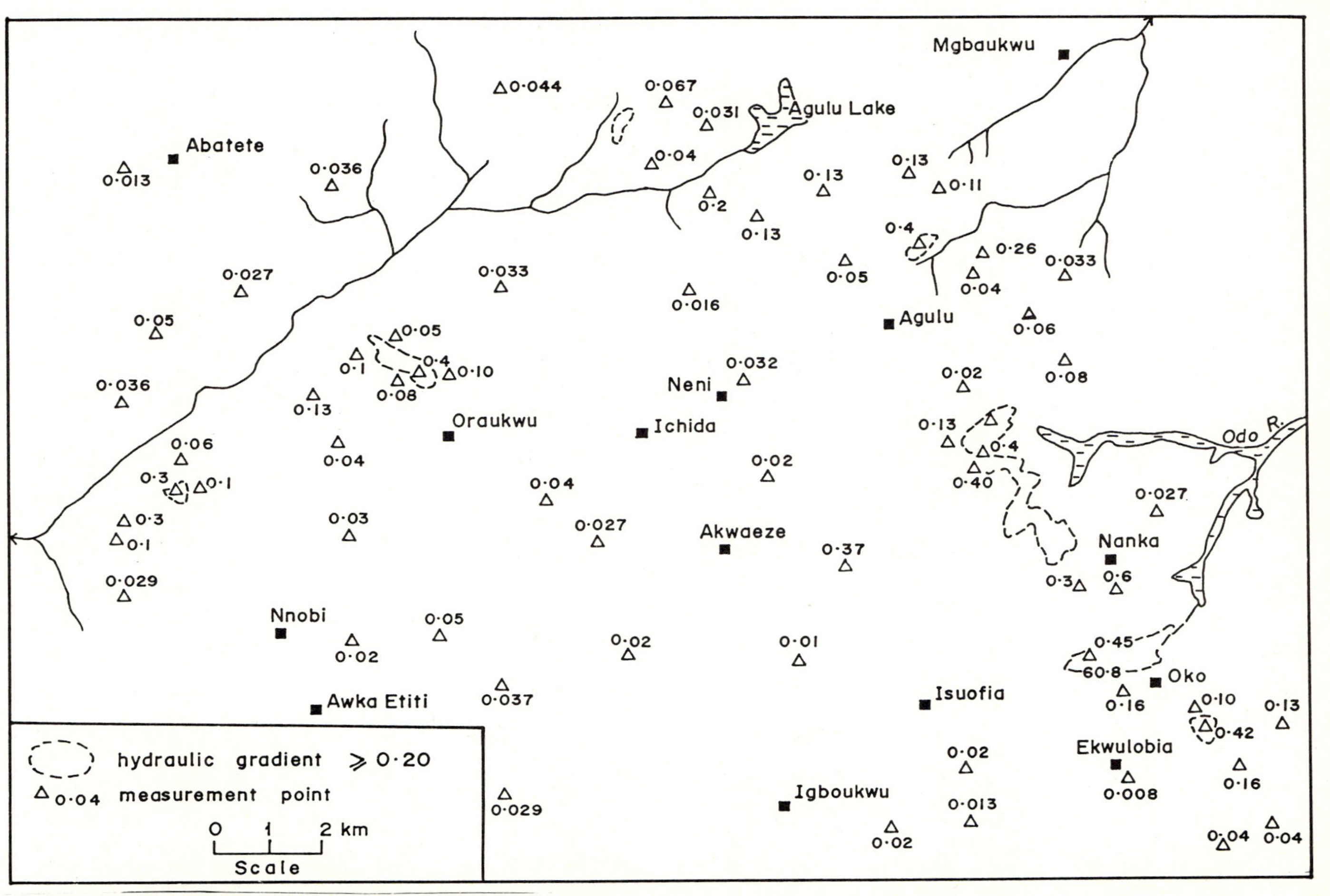

Figure 3. Hydraulic gradients in the Nanka-Oko-Oraukwu area

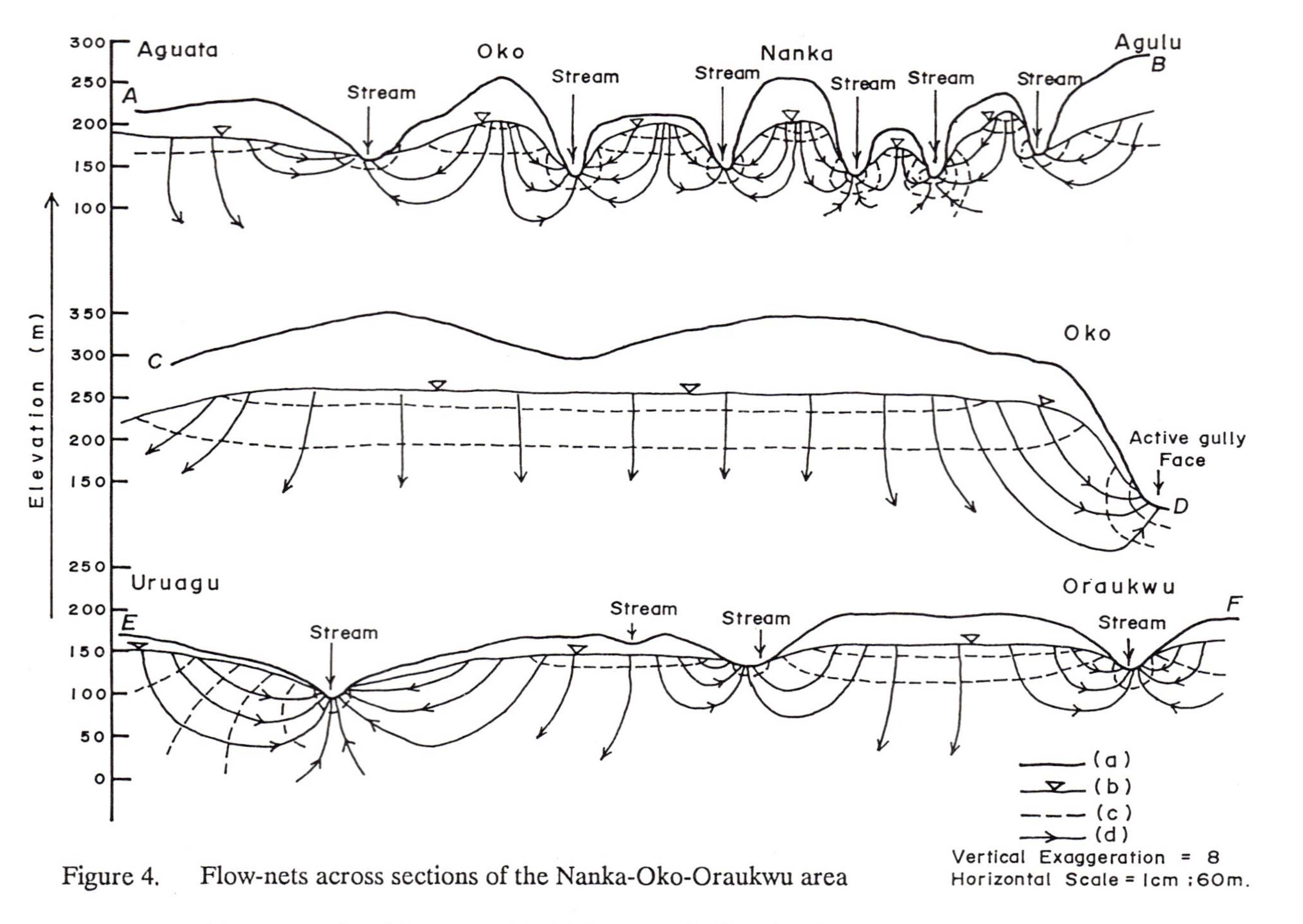

Figure 4. Flow-nets across sections of the Nanka-Oko-Oraukwu area

(a) topography, (b) water table (c) isopotential line (d) flowline

DISCUSSION

The flux of groundwater as shown in Figures 4 and 7 is associated with seepage forces that act in the direction of the flux. According to Cedegreen (1967) the seepage force per unit volume F_x is given as

$$F_x = iM_w \quad (1)$$

where i = hydraulic gradient
M_w = constant = specific weight of water.

From the equation, it can be seen that F_x is directly proportional to the hydraulic gradient. Critical conditions involving quicksand or boiling occur where the hydraulic gradient is about 1.0 (Harr, 1962). In sandy cohesionless materials such as those of the study area, Cedegreen (1967) has shown that the critical hydraulic gradient maybe as low as 0.5.

In the recharge areas, away from the gully heads, the seepage force combines with the submerged weight of the soil/geologic material to improve stability. However, in the vicinity of the seepage forces on the banks of the gullies, the flux is upward directed (Figures 4 and 6). Similarily in these areas the seepage force opposes the submerged weight of the soil/geologic material and stability is reduced. The general hydraulic gradient in the stable areas (Isuochi, Nnobi, Ichida, etc.) is in the order of 0.008 to 0.04 but this increases to about 0.2 in the vicinity of the gullies. In the spots where sliding is frequent, the hydraulic gradient is as high as 0.8 (Fig. 3). The widespread observation of boiling conditions in the field indicates that the exit hydraulic gradient has in many places reached and exceeded the critical value of 1.0. Based on equation 1 and the calculated values of hydraulic gradient which vary from 0.008 to 0.8, the corresponding seepage forces acting on the soil/geological materials have increased 10 to 100 times from the stable areas to the active gully areas. The upward-acting seepage force adds to the activating moment and results in tremendous reduction of the factor of safety of the gully slopes.

Futher more, it has been observed that infiltration from rainfall commonly results in differential increase in hydraulic potential as shown in Figure 8. This gives rise to increases in hydraulic gradient and the corresponding seepage force. The increase in gradient is given from the relation

$$i_2 = i_1 + (h_2 - h_1)/x \quad (2)$$

where i_2 = new hydraulic gradient after infiltration event
i_1 = initial hydraulic gradient
$h_2 - h_1$ = rise in differential hydraulic head between points A and B
x = lenght AB = lenght of soil element under the seepage force.

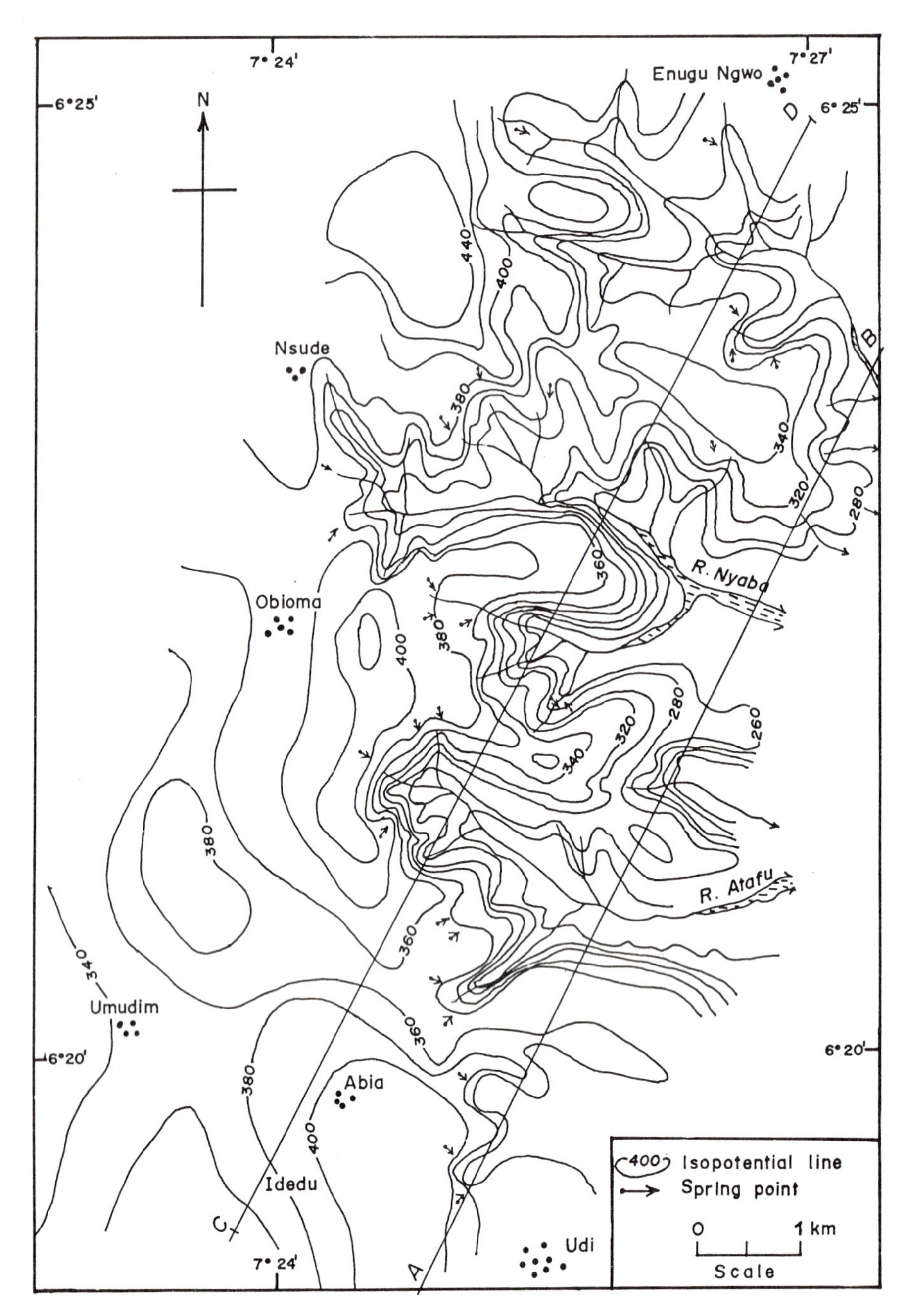

Figure 5. Isopotential map of the Enugu-Udi area

Based on average initial exit hydraulic gradient of 0.5 that is common in the active gully areas, and a theoreticial critical gradient of 1.0, it was found that a rise in differential head of 1.0 m would put 2 m of the soil under boiling or quick conditions. Rainfall in the area commonly results in differential heads that range up to 6.0 m. Under this condition, extensive quick conditions occur and the resulting internal erosion is intensive. The toe of the gully banks are undermined and landsliding and slumping occur.

The water table is high on the slopes. Figure 8 shows the characteristics of one of the active gully spots of Nanka as observed in the field. Using the Odo River as a datum level, it could be seen that the water level, and thus the pore water pressure, is very high (the actual pore pressure distribution could be obtained from closely spaced piezometers or from analyses using numerical methods.) Landsliding of the type that carved out the gully commonly cuts deep below the water table. Failure may be triggered by increasing pore water pressure that reduces, beyond marginal levels, the shearing resistance of the geologic/soil materials which has already been considerably diminished by the seepage force. Such increases in pore pressure are due to percolation of rainwater in the vicinity of the gully slopes.

To obtain a more quantitive impression of the critical role of groundwater, we shall consider a typical gully slope of 45° to the horizontial. The hydrological characteristics of the slope are assumed as in Figure 9 (slope B). The sand materials are cohesionless and have an average specific gravity of 2.63 (Technosynesis, 1978). A soil element of unit dimensions under this condition has a submerged unit weight W_o = 15.99N and the seepage force acting in the direction of the slope is $F_x = M_w i$ = 9.81N. The resultant body force R_c of the seepage force and the submerged unit weight (determined graphically; Cedegreen, 1967) is 24.53N. The resultant body force R_c produces a tangential component T_c = 21.48N and a normal component N_c = 11.77N. Consequently, the factor of safety against failure is F.S. = N_c/T_c = 0.55. The equivalent F.S. for the dry slope is 1.10. The slope under dry conditions would stand (though marginally), but would fail under fully drained condition. According to Cedegreen (1967), the reduction in F.S. is normally around 50% of the F.S. of the dry slope. For gullies with already low factor of safety, as in the example, the reduction in the safety factor introduces super critical conditions and triggers off failure. The assumption that groundwater flux is parallel to the slope is consistent with most egineering analyses but leads to overestimation of the factor of saftey. Under natural conditions, groundwater flux is as shown in Figures 4 and 7 and the contribution of the seepage force to slope instability is more critical. The example used in the illustration represents extreme groundwater conditions and is common on the lower and middle reaches of the gully banks. Full saturation is due to the percolation of rainwater during storms. This type of failure undermines the upper slope (slope A) which eventually collapses. The mode of failure may be soil creep on the fully drained slopes and landslides on the partially drained slopes. In slopes where the water table is not on the surface, the effects of the seepage force is less pronounced and depend on the hydraulic gradient and direction of groundwater

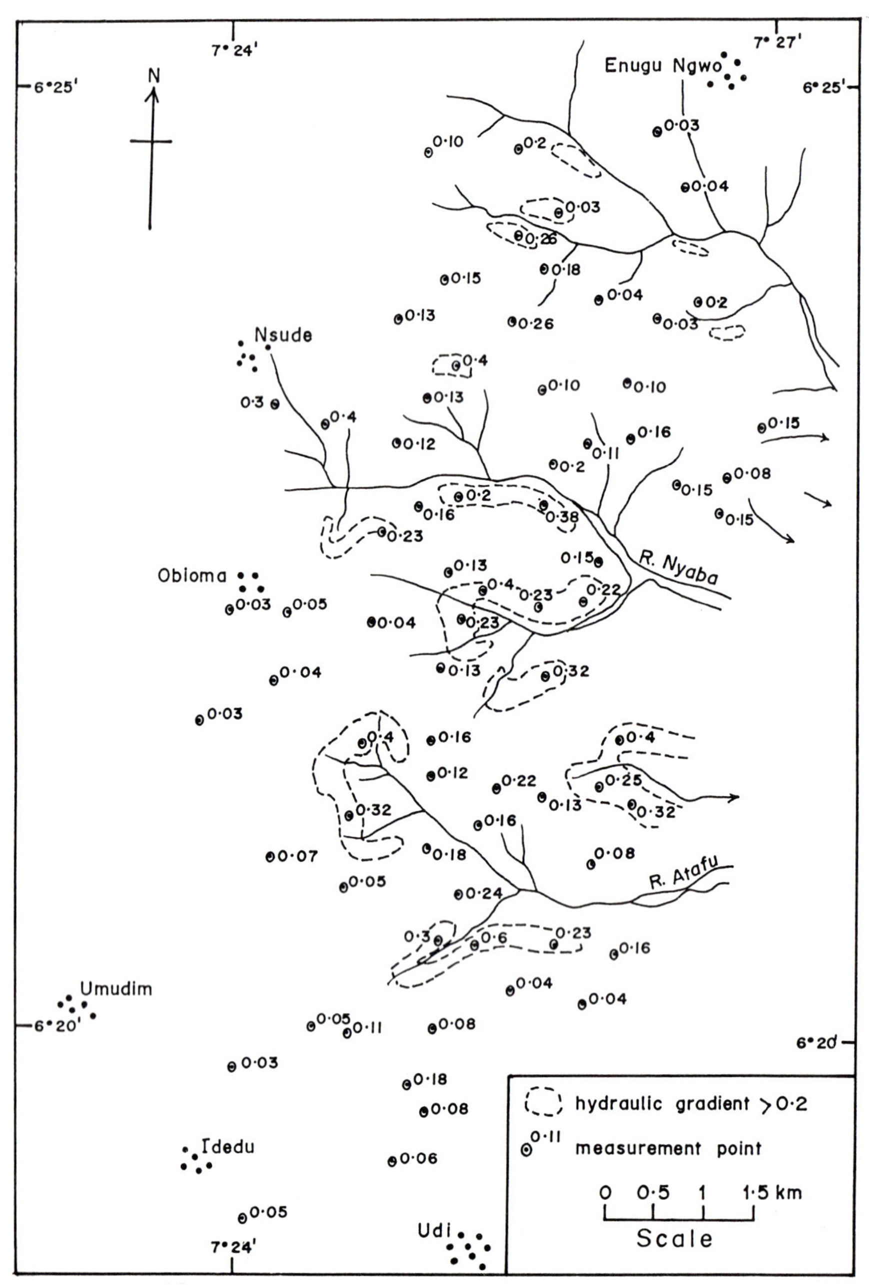

Figure 6. Hydraulic gradients in the Enugu-Udi area

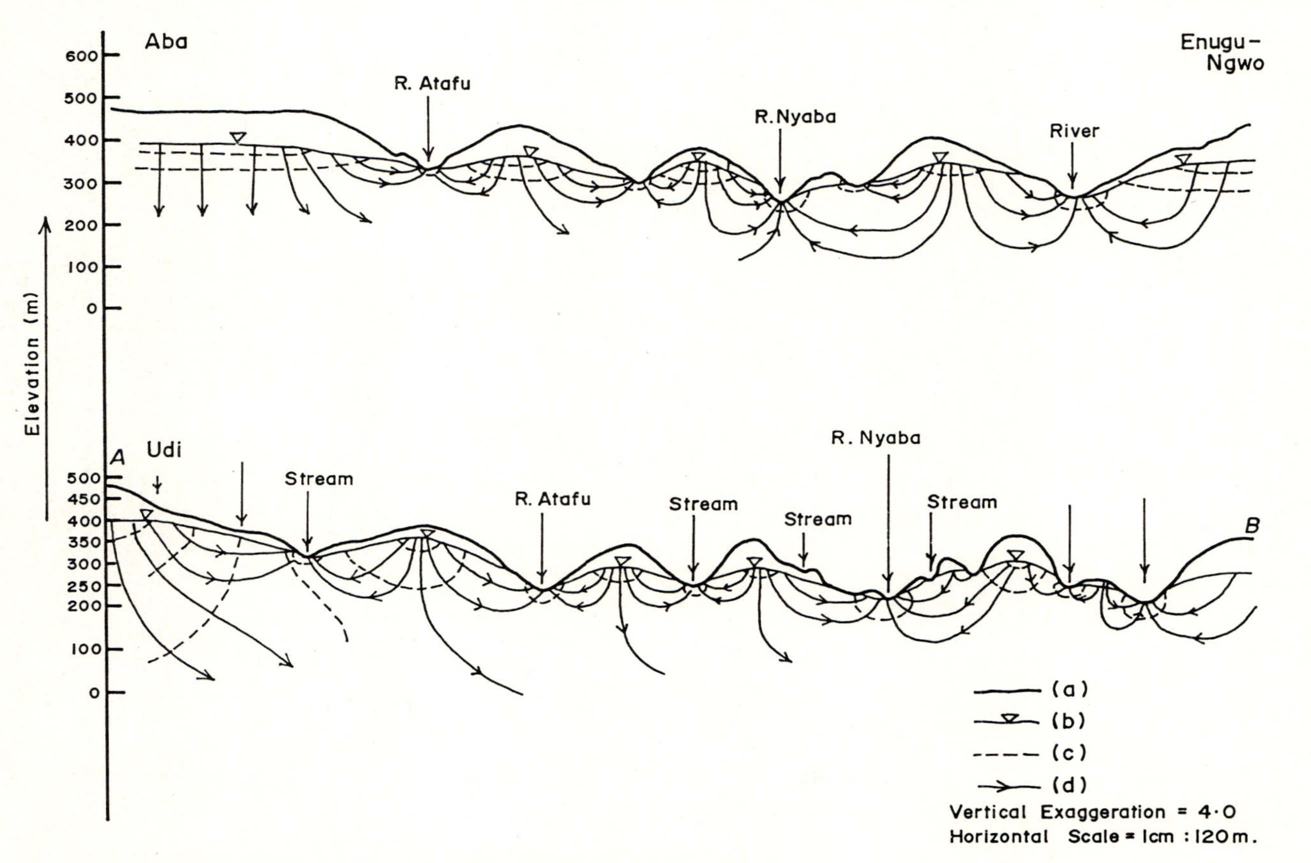

Figure 7. Flow-nets across sections of the Enugu-Udi area
(a) topography (b) water table (c) isopotential line (d) flowline

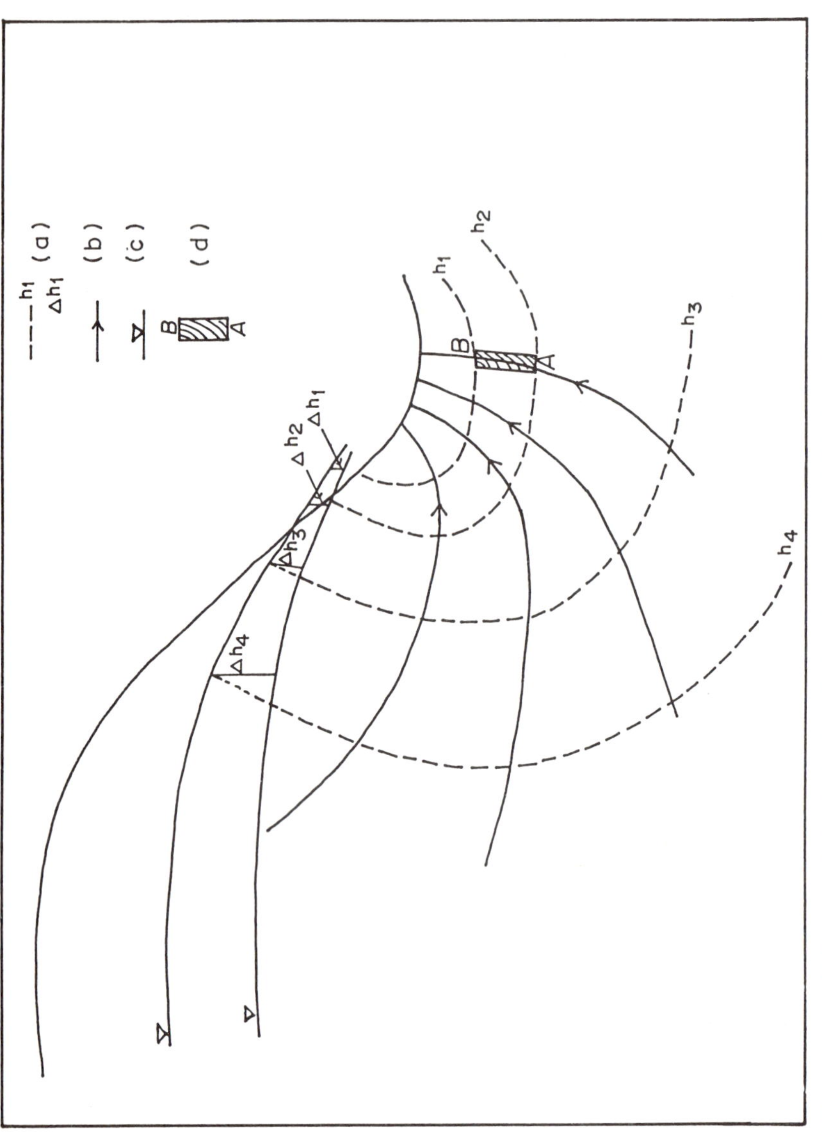

Figure 8. Differential rise in hydraulic head due to rainfall infiltration in a gully slope

(a) isopotential line (h = rise in hydraulic head due to rainfall infiltration)
(b) flowline (c) water table (d) soil element experiencing seepage force

flux. In places where flow is in two pronounced directions as in the vicinity of the gully spots (Figures 4 and 7), analyses of the contribution of the seepage force are more complex but commonly lead to more critical roles.

SUMMARY AND CONCLUSION

Three major erosion phenomena, namely classical gully erosion, bank erosion and mass wasting are acting in various combinations in the active gully areas of southeastern Nigeria. The dominant phenomenon varies from place to place. Classical gullying or accelerated erosion is generally concentrated in the shallow gullies with depths of less than 10 m. Bank erosion is active along the semi-mature river banks. In the most dangerous locations, the major erosive phenomenon is mass wasting through landslides, soil creep and slumping. These are results of frequent failure of the gully slopes. The instability of the gully slopes is due to high and upward fluxing groundwater in the vicinity of the gullies. The high fluxes of groundwater develop high seepage forces and pore-water pressure which reduce the factor of safety of the gully slopes below critical levels and cause the eventual collapse of the slopes. It is our conviction that a better understanding of the dynamics of gullying in the area cannot be achieved without taking these factors into. consideration. In designing lasting control measures, it is essential that the gully heads be dewatered and groundwater levels in the vicinity of the gully lowered. Measures that tend to force more infiltration of rainwater around the gully heads such as those currently being applied can only accelerate gully development through greater fluxes of groundwater and should be stopped. Dewatering of the slopes and lowering of groundwater levels could be achieved through the use of horizontial drains, drainage galleries, well point system or intensive groundwater extraction through deeper boreholes. The last method will in addition help to supply potable water to the affected areas.

ACKNOWLEDGEMENT

In the preparation of this paper the authors benefitted a lot from dicussions with Dr. C.O. Okagbue of the Department of Geology, U.N.N., and Dr. B.C.E. Egboka of the Anambra State University of Technology, Awaka Campus.

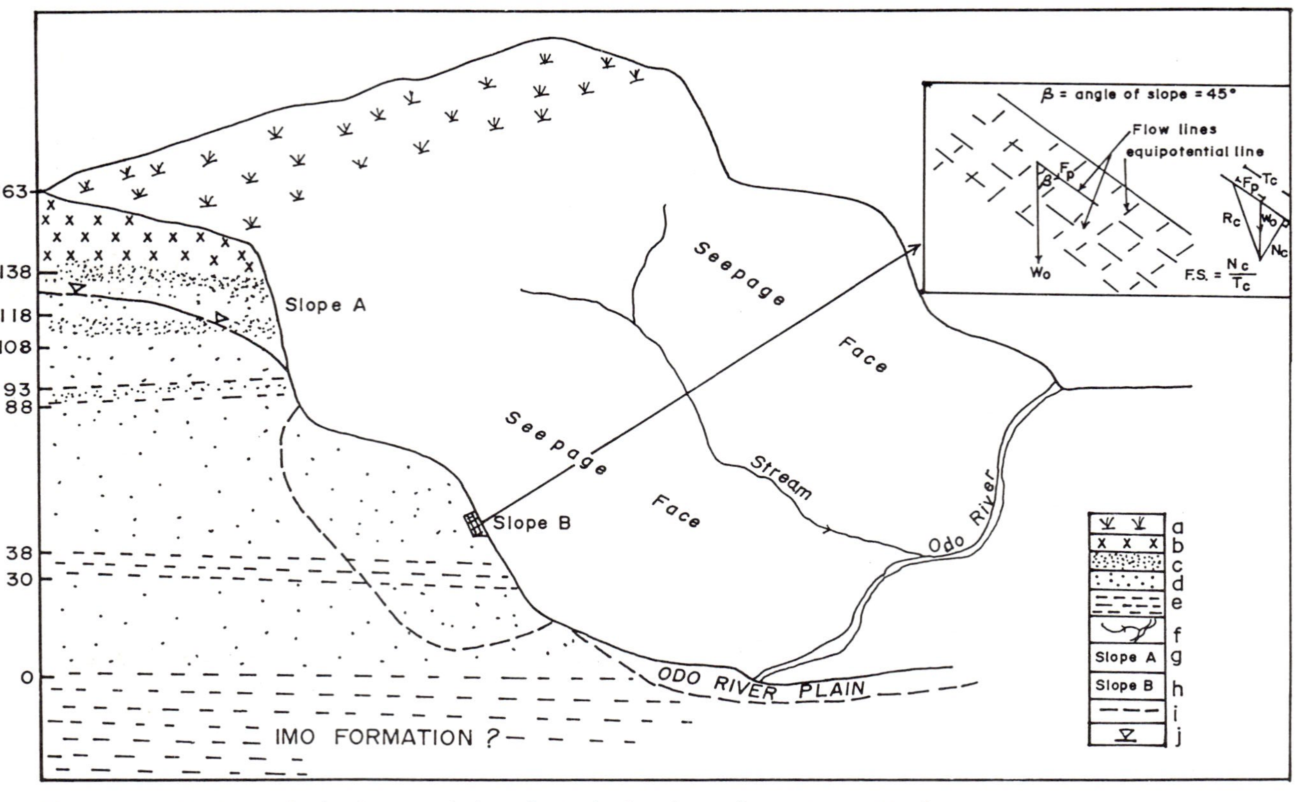

Figure 9. Hydrogeologic characteristics of a typical active gully spot near Nanka

(a) vegetation (b) red earth (c) fine grained sand (d) coarse to medium grained sand (e) clay to sandy clay (f) stream (g) partially drained slope (h) fully drained slope (i) potential failure plane (j) water table

REFERENCES

Akujieze, C.N. and I.N. Ogbukagu, 1985: Hydrogeology of the area (Nanka Sands) south of the upper sector of the Idemili drainage basin. Paper presented at the 21st Annual conference of Nigerian Mining and Geosciences Society Jos, Nigeria.

Cedegreen, H.R. 1967: Seepage, Drainage and Flow Nets. John Wiley and Sons, Inc., New York.

Egboka, B.C.E. and G.I. Nwankwor, 1985: The hydrogeological and geotechnical parameters as agents for gully-type erosion in the Rain Forest Belt of Nigeria.J. Africian Earth Sci.,vol. 3, 417-425.

Floyd, B., 1965: Soil erosion and deterioration in Eastern Nigeria. Nigerian Geogr. Jour., vol. 8, 33-43.

Freeze, R.A. and J.A. Cherry, 1979: Groundwater. Prentice-Hall Inc., Englewood Cliffs, New Jersey.

Harr, M.E., 1962: Groundwater and Seepage. McGraw Hill, New York.

Hodge, A.L.R. and R.A. Freeze, 1977: Groundwater flow systems and slope stability. Canadian Geotech. Jour., vol. 14, 466-476.

Nwajide, C.S., 1977: Sedimentology and Stratigraphy of the Nanka Sand. Unpublished M. Phil. Thesis, University of Nigeria, Nsukka.

Nwajide, C.A., and M. Hoque, 1979: Gullying processes in south-eastern Nigeria. Nigerian Field, vol. XLIV, 64-74.

Ofomata, G.E.K. 1965: Factors of Soil erosion in the Enugu Area of Nigeria. Nigerian Geogr. Jour., vol. 8, 45-59.

Ogbukagu, IK.N. 1976: Soil erosion in the northern part of the Awka-Orlu Uplands, Nigeria. Nig. J. Min. Geol., vol. 13, 6-19.

Simpsoh, A. 1954: The Nigerian Coalfields: the geology of parts of Onitsha, Owerri and Benue Provinces. Geol. Surv. Nigeria Bull., no. 24.

Technosynesis Sp. A., 1978: Soil erosion control in Imo and Anambra States: part 1, control measures against gully erosion. Technical report prepared for the Federal Ministry of Agriculture and Rural Development of Nigeria.

Determination of Polluted Aquifers by Stratigraphically Controlled Bio-chemical Mapping: Example from the Eastern Niger Delta, Nigeria

L. C. Amajor and C. O. Ofoegbu
Faculty of Science, University of Port Harcourt, Port Harcourt, Nigeria

Keywords: Poluted aquifers, Biochemical mapping, Niger Delta, Groundwater quality

ABSTRACT

Seven boreholes, producing about 50,000 gallons of water per hour on the average, supply domestic water to the more than 9,000 inhabitants of the University of Port Harcourt campus and neighbouring inhabitants. Four intergradational and unconfined sand units in the upper section of the Benin Formation (Miocene-Recent) constitute the aquifers (aquifers 1, 2, 3 and 4).

Although the water quality compares favourably well with WHO (1971) international standard for drinking water, a few aquifers are probably being contaminated presently. Biological analysis suggest the presence of coliform organisms (18/100 ml) in aquifer 3 (boreholes III and IV). This is thought to be caused by the nearby polluted New Calabar River which has an influent relationship with the groundwater system in the dry season. On this basis, it is suggested that bio-chemical analysis of the waters be conducted every dry season to monitor the rate and nature of contamination so as to locate its sources and check them.

Boreholes II and IV, located in aquifer 2 produce permanent moderately hard water high in total iron content. The borehole waters are, also, generally weakly acidic and the discharge of organic wastes in the recharge areas of the aquifers are thought to be responsible. Lime or sodium carbonate treatment is recommended to accommodate the possible damages associated with corrosive waters.

INTRODUCTION

The University of Port Harcourt, located in Choba, Rivers State, is about 13 km northwest of Port Harcourt (Fig. 1). It is situated at the junction between the East-West and Rumuokuta-Choba roads. The campus, which is more than 18 sq km is bound by latitudes 4° 54' and 4° 56' north and longitudes 6° 54' and 6° 59' east.

The area falls within the subequatorial region characterized by an average temperature of 25.5^{o} C, a relatively high humidity which fluctuates in the day from 60 % to 95 % and a wet (April to October) and dry (November to April) seasons during the year. Maximum annual rainfall averages 2500 mm. The freshwater swamp vegetation belt characterized by a thick and dense forest formerly engulfed the campus. Today, however, the vegetation has been reduced to grasses by human activities with only a few palm trees and shurbs spotly distributed in the area.

The only surface water in the area is the northeast-southwest flowing New Calabar River which currently forms the western boundary of the campus. The river is

Figure 1
Location of study area showing rough hydrogeological zones in the eastern Niger delta. a) shallow sand and gravel water aquifer; b) shallow to deep sand aquifer mixed with clay; c) deep sand aquifers.

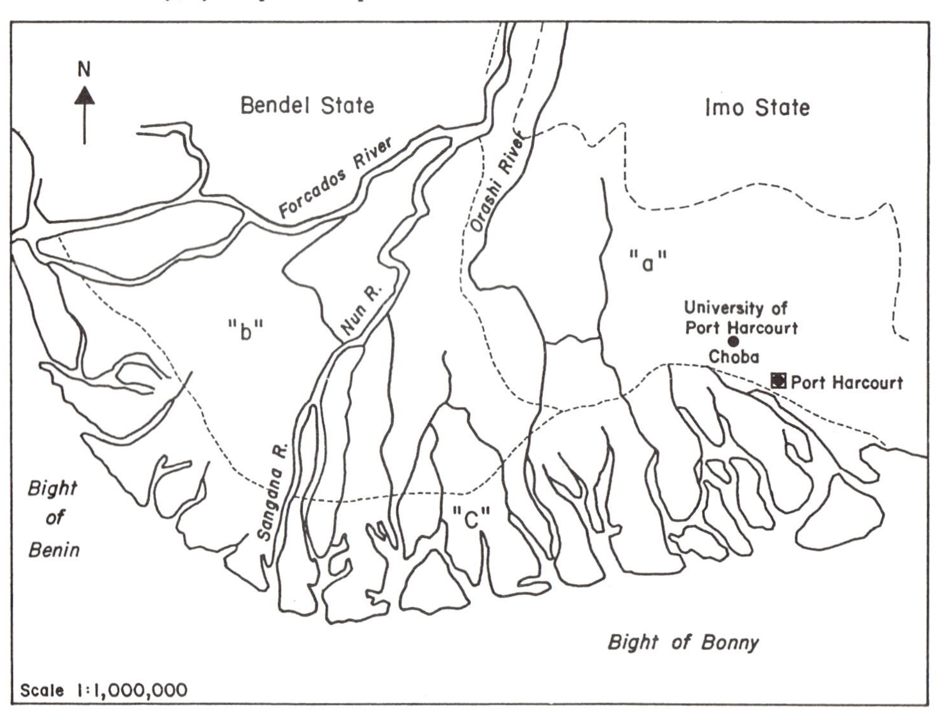

polluted with industrial and human wastes from nearby industries and villages and as such is unsafe for domestic use and human consumption. Thus, the University turned to groundwater as the main source of supply of adequate potable water to her ever increasing population. Between 1976 and 1983, seven water boreholes have been completed, but five are currently producing. These yield about 8,000 to 10,2000 gallons of water per hour.

As the population of the area continues to grow at a rapid rate, water yield from existing boreholes is likely to decrease and more boreholes may fail due to over pumping. The need to maintain continues water supply will subsequently lead to the construction of more boreholes. The location of such boreholes and the aquifers they should tap should be a direct function of the hydrogeology of the area, rather than population density and proximity as had been the case with earlier boreholes. This, it is believed, will lead to a better borehole construction and management programs.

The essence of this paper is to document, under one cover, the known hydrogeology of Choba area based on the analysis and interpretation of the discrete subsurface data retrieved from existing boreholes and compiled by the different borehole drilling contractors.

STRATIGRAPHY

The study area is within the lower section of the upper flood plain deposits of the subaerial Niger Delta (Allen, 1965). The deposits are characterized by pebbles and coarse to fine sands with intercalations of silt, mud and clay in places. A variety of depositional environments (point bars, channel fills, natural levees and splay deposits, backswamps, ox-bow fills and palludal deposits) are typical (Nedeco, 1959; Allen, 1965).

Underlying these Quaternary sediments is the Benin Formation which is about 2100 m thick on the average (Short and Stauble, 1967). The formation represents the subsurface continental megafacies of the Niger deltaic sequence. It is essentially fluvial in origin and comprises of unconsolidated, massive, and porous freshwater bearing sands with localized shale interbeds. All the aquifers in the delta region are located within this lithounit. However, only the upper 130 m has been penetrated by boreholes in the study area. The age ranges from Miocene to Recent.

The underlying paralic Agbada Formation, varies in thickness from 300 to 4500 m. The formation consists predominantly of unconsolidated pebbles, and very coarse to fine grained sand units with surbordinate shale beds. Syntectonic growth faults and rollover anticlinal structures are characteristic. These commonly form the major traps for oil and gas in the delta. Age ranges from Oligocene to Recent.

ELEMENTS OF THE HYDROLOGIC CYCLE

According to G.R.B.M. (Nig.) Ltd. (1979), the mean annual precipitation, evapotranspiration and run-off in the region are, respectively estimated at 3000 mm, 1143 and 1200 mm. These leave us with an annual water budget of 657 mm. This figure indicates that a fairly large quantity of water percolates underground every year. However, the lack of information on the recharge and discharge characteristics for the various aquifers makes it difficult to advice on the pumping rate, exact or otherwise, for any well.

The elevation of the water table fluctuates between 2 and 5 m, while that of the surface of Calabar river stays between 6 and 9 m. These variations are to be expected as they reflect changes in the climatic pattern of the area.

During the wet season the water table and the river surface are at about the same elevation (6 m). Under this condition the river is neither influent nor effluent. Since the river water is normally more turbid at this time of the year it may infiltrate into the adjacent soil probably because of its greater relative density. A tendency towards influent conditions is likely (Fig. 2a) in the dry season, however, the water table drops below the river level. A difference in elevation is created and the river becomes influent and recharges the aquifer. In both cases, there is the danger of the river water contaminating the groundwater especially in the dry season (Fig.. 2b). Should this occur, the aquifers being tapped by boreholes 3 and 4 at Delta Park will be affected first. Thus, there is the urgent need to constantly monitor the chemistry of the borehole water every dry season.

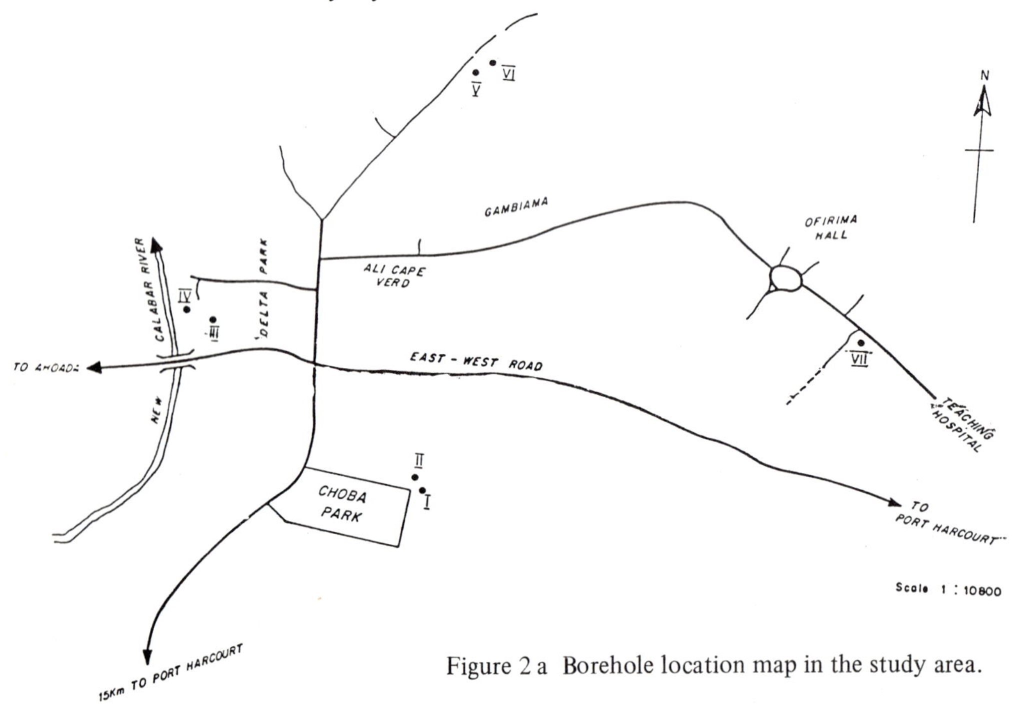

Figure 2 a Borehole location map in the study area.

Figure 2 b Schematic west-east section showing the relationship between groundwater and the new Calabar River in the dry season

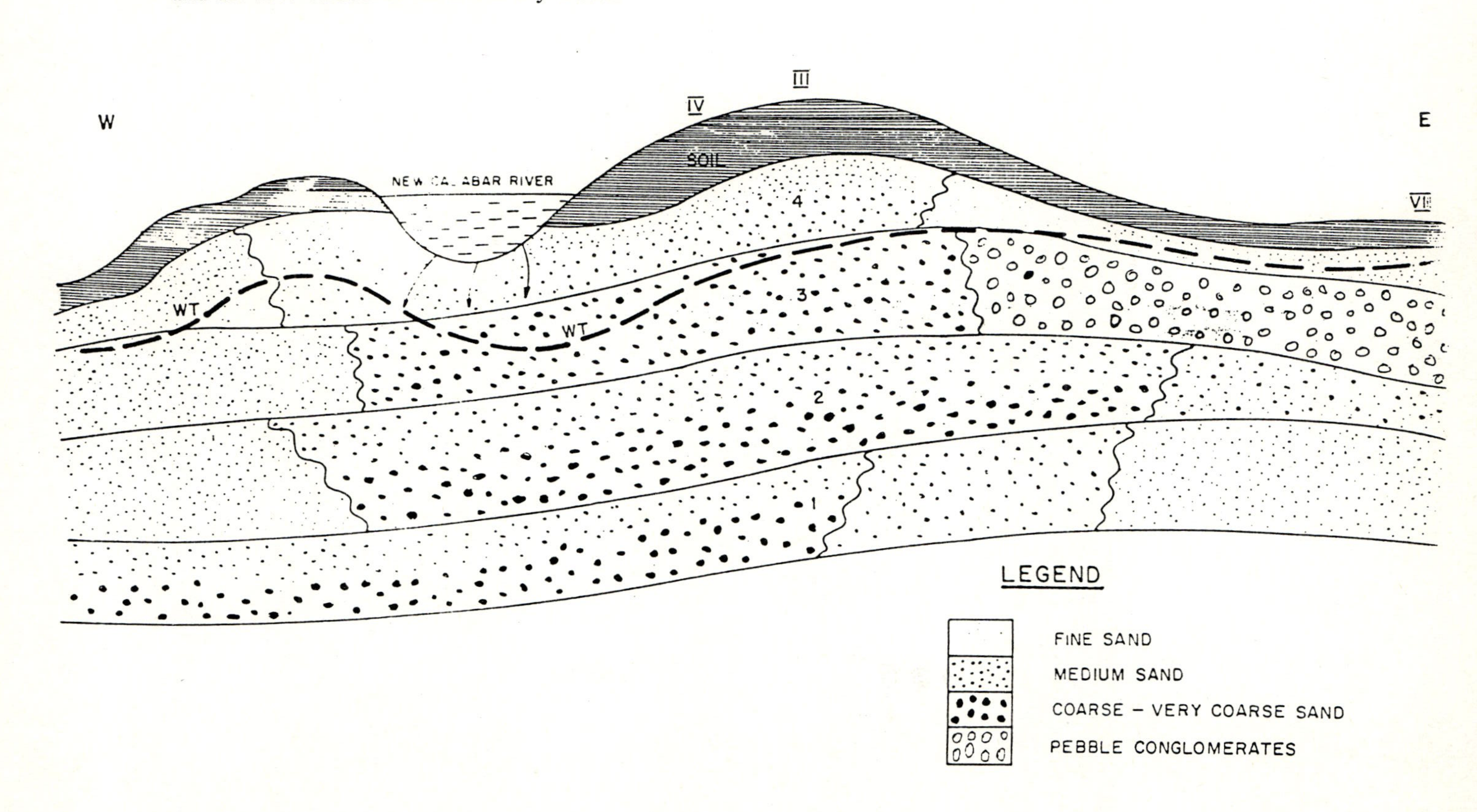

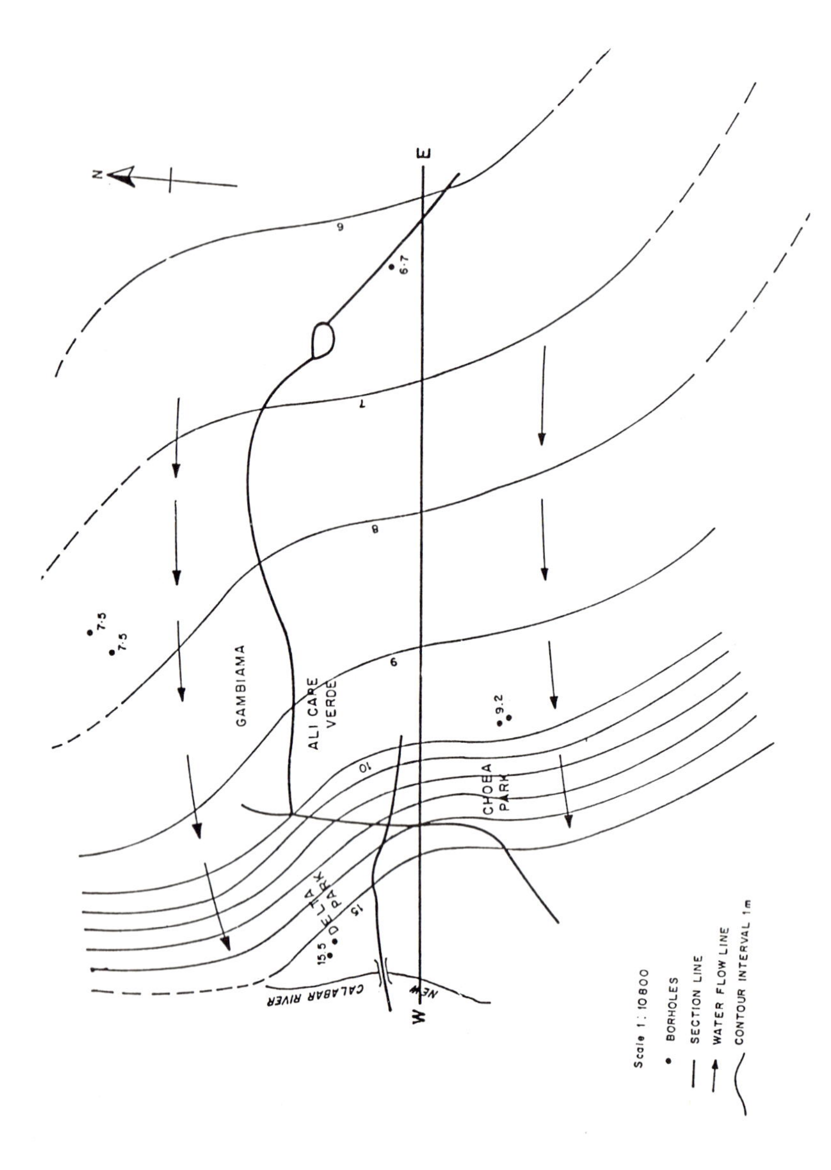

Figure 3 Water table map of the study area.

On the basis of the water table map (Fig. 3), the regional fow pattern of the groundwater appears to be from the north-east towards the southwest.

AQUIFER CHARACTERISTICS

Figure 4 is a three-dimensional configuration of the stratigraphic succession of the study area. It begins with more than 85 m thick sequence of conglomerates and sand beds of variable texture. These constitute the aquifers and about 2 m thick reddish sandy clay soil caps the sequence. In general, silt, mud and clay intercalations seem absent. This stratigraphic succession indicates that the aquifers are unconfined.

Four aquifers, corresponding to four cyclic fluvial sequences without intervening flood plain clay deposits, were recognized within the drilled portion of the Benin Formation based on textural characteristics (fining upward sequence) and lateral facies changes. Aquifer 1, which occurs at the base within a depth range of 86 to 130 m, is partly encountered in boreholes II, IV and V. It consists of coarse to medium sand to the southwest (wells II and IV) becoming finer grained to the northeast (wells VI and VII). No boreholes tap water from this aquifer except the failed ones (III and V) for which there is no data. Aquifer 2 was penetrated by all the boreholes, except VII, between 73 to 93 m depth. Very coarse to coarse grained sand in the west which grades laterally into finer sands to the east characterize this litho-unit. Boreholes II and VI produce water from this aquifer. Aquifer 3, which occurs between 33 to 73 m depth, was penetrated by all the seven boreholes. It is composed of pebble conglomerates in the east and these grade into coarse sand to the west. Boreholes VII produces from the conglomerate facies while I and II tap the waters of the coarse sand units. Aquifer 4, which is the highest unit, lies within 2 - 33 m depth range. It comprises coarse and medium to fine grained sands. No borehole produces from this sand unit.

In general, a fining upward textural gradient typical of luvial sequences characterize all the aquifers. The pebbles and very coarse to fine grained sand are angular to subrounded and poorly to fairly well sorted in places. Figure 5 shows the grain size distribution of aquifer 2 from water boreholes II and VI. The curves suggest a fairly uniform sand size range comprising mostly monogranular coarse sand (81 - 85 %) and pebbles (15 - 19 %).

On the basis of groundwater temperature of 24° C, viscosity of 1.308 at 10° C, and a specific gravity of 0.896 at 24° C, the computed hydraulic conductivities (field coefficient of permeability) range from 3.9×10^{-2} cm/s for well II, to 4.1×10^{-2} cm/s for well VI. These figures indicate that aquifer 2 is very permeable, yielding 8,000 to 10,000 gallons per hour. The yield of aquifer 3 (boreholes I, IV, VIII) which varies from 10,000 to 10,100 gallons per hour indicate that its of very good permeable character.

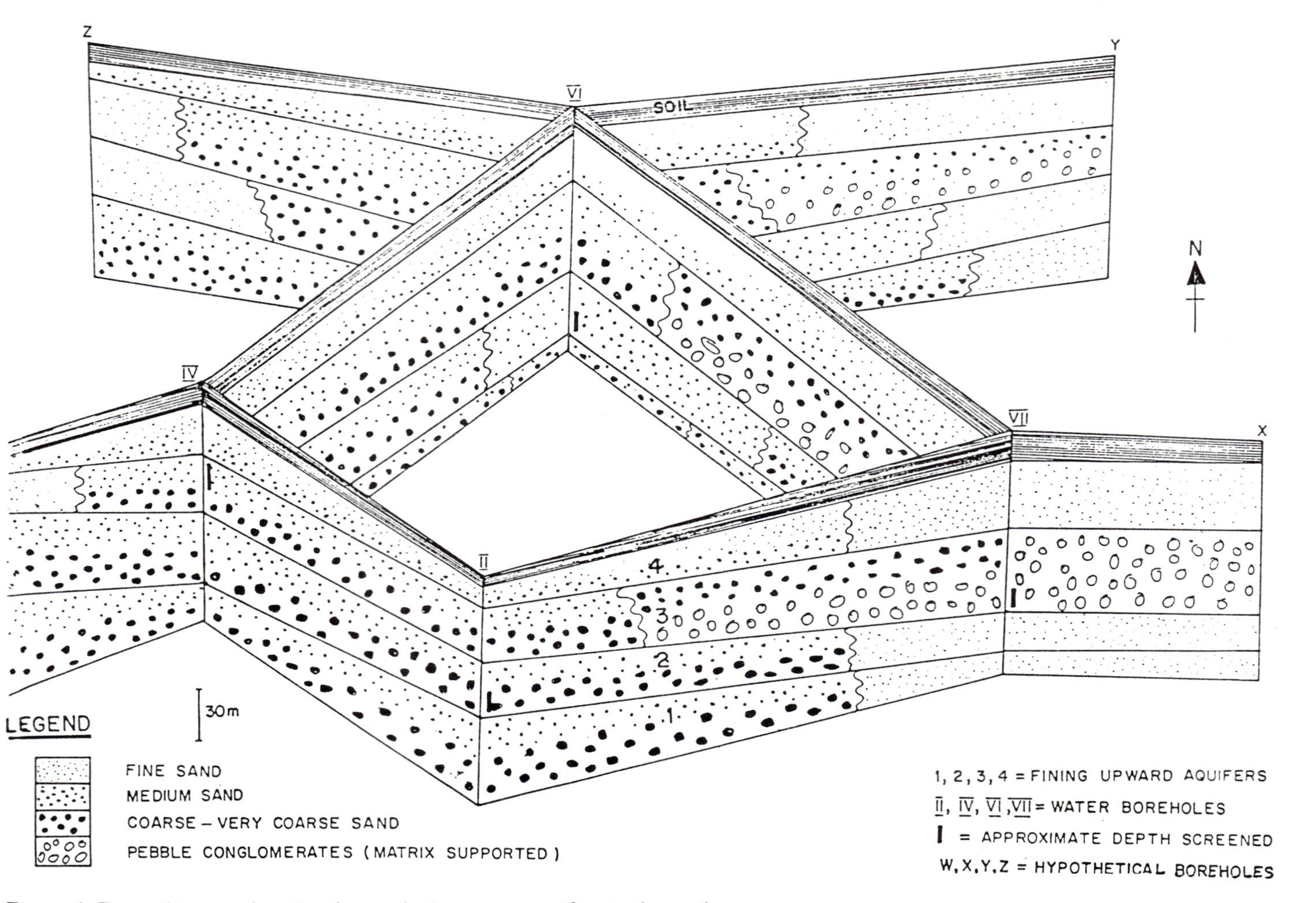

Figure 4 Fence diagram showing the producing water aquifers in the study area.

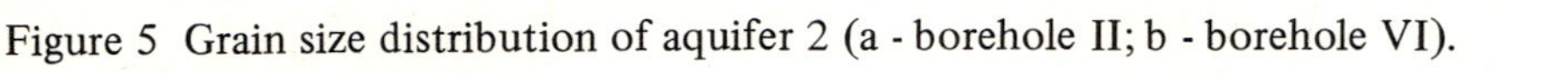
Figure 5 Grain size distribution of aquifer 2 (a - borehole II; b - borehole VI).

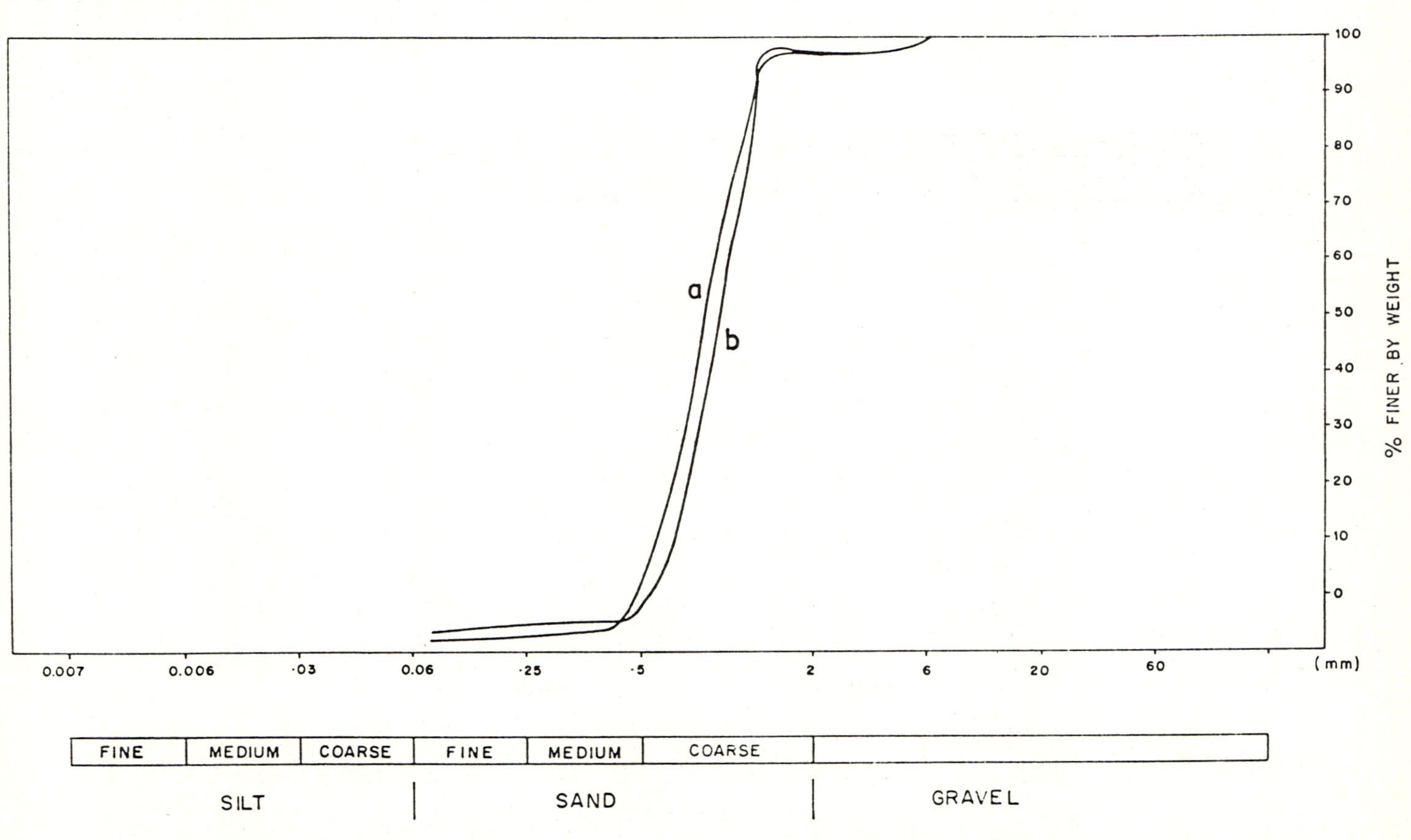

GROUNDWATER QUALITY

The analysed physico-bio-chemical parameters of the groundwater in comparison to that of WHO (1971) international standard for drinking water is shown in Table 1.

The water in all the aquifers is generally clear or colorless (5 HU). The exception is borehole VII which produces slightly clear (34 HU) water. This borehole taps water from the conglomerate facies of aquifer 3. It is most likely that the conglomerate has a matrix support fabric with abundant finer grained sediments as matrix materials. Continuous pumping forces these finer materials into the well thereby slightly discloring the water. The high total suspended solids (45 ppm) in this water lends credence to this interpretation. Although this level is tolerable it could constitute maintainance hazard. For instance a well discharging about 900 m^3 of water per day with 10 ppm total suspended solids will produce 3.32 tones of solids per annum. If not removed, these may lodge in pipes and tanks of the distribution system, and thereby greatly incrase mainteinance cost.

Biologicial analysis shows that most aquifer waters are free of coliform organisims except boreholes III and IV in Delta Park with a high content of 18 per 100 ml (Okoli, 1983). This is higher than the recommended tolerable level of less than 10 per 100 ml. Moreover, the high saline amonia content of wells I, IV and VII, which are 23, 19 and 56 ppm, respectively, exceed the tolerable level of 15 ppm. This suggests that the groundwater system is highly susceptible to bacteriological contamination from human and animal wastes. These bacteriological characteristics of our groundwaters could constitute a major health hazard, if unchecked, because of the possibility of infection.

The writers attribute the source of the coliform, organisims in wells III and IV to the polluted Calabar River whose obvious influent characteristics in the dry season are likely to introduce these organisims into the groundwater regime (see Fig. 2). The suck-away pits distributed here and there in the campus, and the 3 to 4 m deep waste pit behind student hostels in Delta Park are the most likely sources of the saline amonia and coliform organisims too. Mechanical infiltration, and hydrodynamic dispersion are some of the carrier processes likely to introduce the bacteria into the groundwater system.

If these analysis are correct (i.e. coliforms present in the groundwater system) something should be done immediately to rectify the situation before much harm is inflicted on the users of this water. We recommend the following:

a) Re-analysis of all waters from the boreholes, preferably different laboratories should be used so as to check the accuracy of the results.

Table 1 Some groundwater quality parameters in the eastern Niger delta compared with the WHO (1971) International Standard for drinking water.

	BOREHOLES							
PARAMETERS (ppm)	I	II	III	IV	V	VI	VII	WHO(1971) ma/1
Appearance	Clear	Clear	Clear	Clear		Clear	Moderately clear	
Magnesium	7		4	8				30 150
Potassium								
Iron (total)	0.04	0.61	0.04	0.04		6.01	0.01	0.1 1.0
Manganese								0.05 0.5
Zinc								5 15
Copper								0.05 5
Chromium (6+)								- 0.05
Silica	16	10.43	18	14		11	32	
Arsenic								- 0.05
Cadmium								- 0.01
Cyanide								- 0.05
Lead								- 0.01
Mercury								- 0.001
Selenium								- 0.01
								-
Phenon								0.001 0.002
Anionic Detergent								0.02 1.0
Polynuclear Aromatic Hydrocarbons								0.000
Gross Alpha Radio-activity (PC/L								3
Gross Beta Radio-activity (PC/1)								
E. Coli in 100ml								0
Coliform organism in 100ml								

Table 1 (continued overleaf)

Table 1 (continued)

	BOREHOLES							WHO (1971) DRINKING WATER STANDARD
PARAMETERS (ppm)	I	II	III	IV	V	VI	VII	WHO(1971)mo/
Appearance	Clear	Clear	Clear	Clear		Clear	Moderate Clear	
Color(Hazen Units	5	5	5	5		5	34	5-56
Total suspended solids	12	20.1	12	26		21	45	
Conductivity (Uhmos)	HD	16	ND	ND		16	38	
Total Dissolved Salts	20	14	ND	40			14	25
Total Solids	32	34.14	62	66		35	70	500-1500
pH	6	6.2	6.6	5.65		6	6	6.5 9.2
Acidity								
Total Alkalinity	27	2.15		25		2	1	
Calcium Hardness	28			20				
Magnesium "	7			8				
Total Hardness	35	78.8	50	28		80	20	100-500
Chloride	5.05	20.51	ND	4.84		19.74	15.03	200-600
Sulfate	18	7.2	21	18		7	8	200-400
Nitrate	0.01	0.01	0.01	0.01		0.01	0.01	45
Phosphate								
Carbonate								
Bi-Carbonate	33	2.0	58	21		2.4	1.22	
Fluoride	0	0.01	0	0		0.1	0.2	0.6 0.7
Saline Ammonia	23	2.01	15	19		2	56	15ppm
Dissolved Oxygen								
Sodium	34	48	38					
Calcium	28	46	20				75	200

b) Most of the disposal pits may have to be emptied, backfilled and properly relocated and constructed based on the hydrogeologic characteristics of the place.

c) Treatment for bacteria.

Within the limits of the chemical analysis, the groundwater system of the University of Port Harcourt area could be said to be chemically fit for drinking and other domestic purposes.However, a few anomalies are disernable. Generally the water is very weakly acidic (pH 5.65 to 6.99), while aquifer 2 (boreholes II and VI) produces permanently moderately hard (78.8, 80 ppm) water high in the total iron content (0.61 and 6.01 ppm) Ugboaja (1983).

The above chemical anomalies may lead to corrosion of the iron and steel materials (casings, pipes and plumbing fixtures), incrustation or clogging of well screens and distribution pipes, cause objectionable taste to drinks and foods, may stain clothes and rust cooking utensils. In addition iron bearing waters favour the growth of iron bacteria eg. Crenothrix which exerts a marked clogging effect and reduces flow rate.

In borehole I,located at Choba Park, the submersible pump dropped right aside the screen, and has since not been picked up. Although the cause is not too well known, borehole contractors, who could not fish out the pump, attribute it to corrosion.

CONCLUSION

1. This study shows that of the 7 boreholes constucted to supply domestic water to the University of Port Harcourt Community only 5 are functioning at present while 2 are bad due to unknown causes as yet.

2. These produce about 50,000 gallons of water per hour from three sharply gradational sandstone aquifers.

3. The aquifers are generally unconsolidate, friable, pebble to fine sand grained in size, subangular to subrounded and poorly to fairly sorted: Each aquifer is about 25 m thick and exhibit a fining textural gradient interpreted to reflect a fluvial depositonal system.

4. The hydraulic conductivities of the aquifers are high and the watertable elevation varies between 6 and 15 m. The groundwater flow pattern is from east and/or northeast to west and/or southwest.

5. The relatively high total suspended solids observed in well VI could be due to the fine sediment matrix that support the pebble conglomerate fabric from where the borehole produces.

6. Boreholes III and IV which tap water from aquifer 3 contain coliform organisims (18/100 ml.) This is thought to be due to the influent relationship between the groundwater system and the polluted New Calabar River.

7. The water is generally weakly acidic and boreholes II and VII situated in aquifer 2 produce hard water high in total iron content (0.61 - 6.2 ppm.). Discharge of organic wastes in the recharge areas of the aquifers are thought to be responsible.

8. It is recommended that lime or sodium carbonate treatment be carried out to improve its physico-chemical characteristics and thereby or eliminate damages associated with corrosive waters. Bio-chemical analysis of the waters should be conducted every dry season to monitor the rate and nature of contaminants so as to delineate its source and eradicate it. Moreover, the water analysis should be more detailed to include the toxic elements.

9. In general, the locations of future boreholes should now be guided by hydrogeology rather than population density.

ACKNOWLEDGMENT

This field assistance of Okoli M. and Ugboajah R. and Laboratory assistance of Rivers State Utliity Board are acknowledged.

REFERENCES

Allen, J.R.L., 1965: Late Quaternary Niger delta and adjecent areas: Sedimentary environments and lithofacies. American Assoc. Petroleum Geologists Bull. Vol. 49, 549-600.

G.R.B.M. (Nig.) Ltd., 1979: Hydrogeological investigations in the sedimentary basins (Coastal Plains and Niger Delta), Vol. I, Unpubl. Rept. Fed. Min. Water resources, 210.

Nedeco (Netherlands Engineering Consultants), 1959: River Studies and recommendations on improvement of Niger and Benue Rivers, 580.

Okoli, M.N., 1983: Groundwater exploration and exploitation in the University of Port Harcourt. Unpubl. B.Sc. project, UNIPORT., 52.

Short, K.C., and A.J. Stauble, 1967: Outline of the Geology of Niger delta: American Assoc. Petroleum Geologists Bull., Vol. 51, 761-779.

Ugboaja, R.N., 1983: Groundwater survey of University of Port Harcourt. Unpubl. B.Sc. project, UNIPORT, 63.

WHO, 1971: International Standards for Drinking Water, 3rd Ed., Geneva.

Calculating Effective Reservoir Formation Properties in Aquifers or Hydrocarbon Reservoirs having a Sinusoidal Pressure History

E. O. Udegbunam
University of Port Harcourt, P.M.B. 5323, Port Harcourt, Nigeria

Keywords: Formation properties, Aquifers, Hydrocarbon reservoirs, Sinusoidal pressure history, Mathematical model

ABSTRACT

This study involves the formulation and solution of a mathematical model which permits the determination of effective in-situ formation properties in aquifers and hydrocarbon reservoirs using their past sinusoidal pressure history.

Previous reservoir welltest models and techniques have had limited applications in reservoirs with sinusoidal pressure history.

Two applications of the model are shown. The first application validates the model and shows the sensitivity of result(s) to errors in the input data while the other application calculates the hydraulic diffusivity in a natural gas reservoir. Satisfactory agreement was obtained when the pressures at the downstream well location were reconstructed from the model using the calculated hydraulic diffusivity. The latter result shows that the model may be used for pressure prediction in such in-situ systems if the effective formation properties are available.

INTRODUCTION

In-situ formation properties whenever available, represent the effective values and are preferred to laboratory measurements on reservoir rock core samples. In aquifers, especially those being considered for natural gas storage, pumping tests are carried out and analysed for in-situ permeability and compressibility. The Horner's point source solution of the diffusivity equation for fluid flow through the porous media is used in the analysis of the pressure drawdown and builup data (Horner, 1951; ERCB, 1979). In hydrocarbon reservoirs tests like the drill stem test (DST), repeat formation test (RFT), and the long-term production tests (LTT) are also used to obtain drawdown and buildup data and analysed on the basis of the Horner's model (Katz, et al., 1963). The interference and pulse tests, which always involve the simultaneous use of more than a single well in the reservoir, have also been employed in analysis of these formation characteristics (Mueller and Witherspoon, 1965; Johnson et al., 1966).

However, application of these techniques has not been extended to in-situ systems with oscillating and sinusoidal pressure (and/or production) history without the cessation of normal reservoir operations. In In-situ systems having the sinusoidal pressure history include underground gas storage pools with their seasonal gas injection and withdrawal operations. They also include aquifers which surround or are in pressure communication with such gas pools.

The present study attempts to evaluate formation hydraulic diffusivity using the past sinusoidal pressure history of this special class of underground reservoir systems. Two examples are considered. The first one is used to prove the validity of this model and also to reveal the sensitivity of results to any errors in the measured pressure history. The second case analysed past pressure history from a natural gas storage reservoir according to the present model. This technique is equally applicable in water sands (aquifers) and natural gas storage reservoirs.

MATHEMATICAL MODEL

The basic element in this new method is an adjacent well-pair which measures pressures at their respective locations. This is depicted in Figure 1. Other requirements are:

a) the well-pair is in a homogeneous fluid,
b) Pressure communication must exist between the well-pair, and
c) no active well is sandwiched between the well-pair.

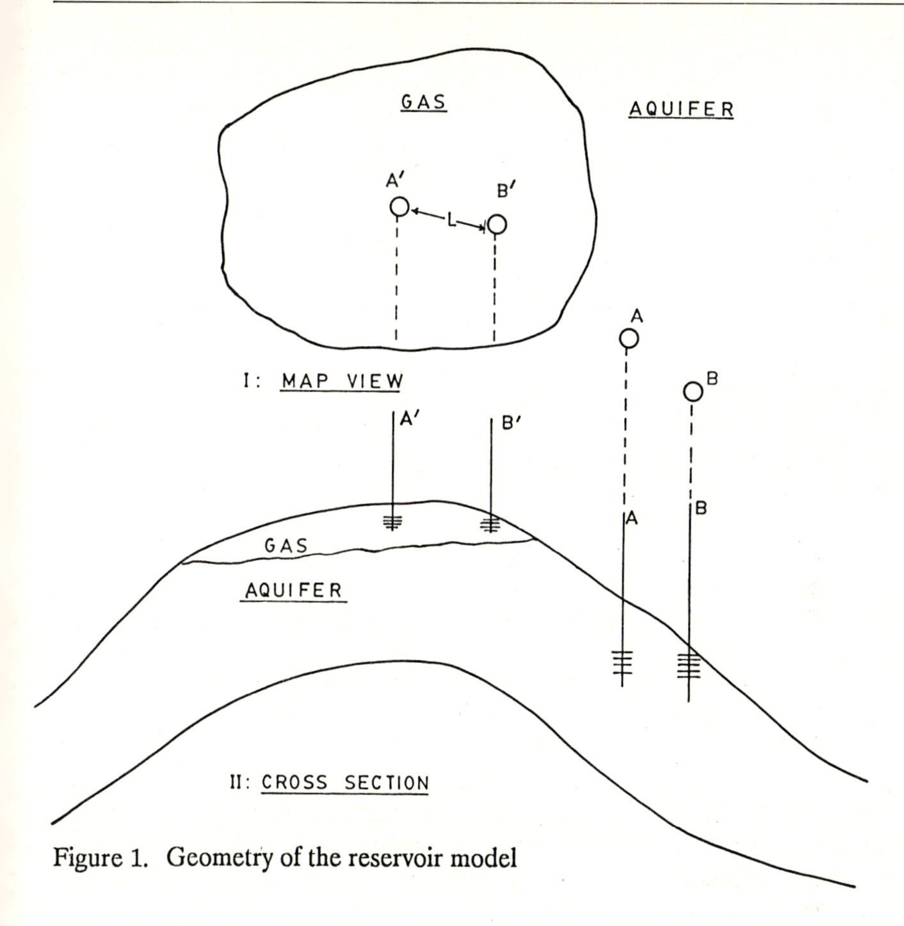

Figure 1. Geometry of the reservoir model

Fluid flow between the wells in the aquifer is governed by the one-dimensional transient flow equation:

$$\alpha \frac{\partial^2 P}{\partial x^2} = \frac{\partial P}{\partial t} \tag{1}$$

When applied to gas, P is replaced by the real gas potential (Φ) which is defined as

$$\Phi = \int_{P_r}^{P} \frac{P\,dP}{\mu z} \tag{2}$$

The well pressure data from the upstream well (forcing function) A and the downstream well (response function) B can be expressed in terms of sinusoidal functions:

Well A: $$P_A = \frac{1}{2} a_0 + \sum_{k=1}^{N} (a_k \cos k\omega t + b_k \sin k\omega t) \tag{3}$$

or: $$P_A = \sum_{-N}^{N} c_k \exp(ik\omega t)$$

Well B: $$P_B = \frac{1}{2} u_0 + \sum_{k+1}^{N} (u_k \cos k\omega t + v_k \sin k\omega t) \tag{4}$$

or:

$$P_B = \sum_{-N}^{N} d_k \exp(ik\omega t)$$

For completeness, let the initial condition between the wells be:

$$P(x,0) = f(x) \tag{5}$$

Equations (3) and (4) are observed pressures at the wells. It has been shown that the initial condition (5) contributes negligibly to the solution at large times (Udegbunam, 1983).

Equations (1) and (3) are satisfied by

$$P(x,t) = \sum_{k=-N}^{N} \exp(-\beta_k x) \cdot c_k \exp\{i(k\omega t - \beta_k x)\} \tag{6}$$

where: $\beta_k = \sqrt{k\omega/2\alpha}$

At the downstream well, the calculated pressure response must be the same as the observed pressure. Hence:

$$\sum_{-N}^{N} d_k \exp(ik\omega t) = \sum_{-N}^{N} \exp(-\beta_k L) \cdot c_k \exp\{i(k\omega t - \beta_k L)\} \tag{7}$$

From (7), we conclude that:

$$\beta_k L = \frac{1}{(1+i)} (\ln \{c_k/d_k\}) \tag{8}$$

Letting $c_k = R_k \exp(i\theta_k)$ and $d_k = r_k \exp(i\sigma_k)$, then

$$\beta_k L = \frac{1}{(1+i)} \{\ln(R_k/r_k) + i(\theta_k - \sigma_k)\} \tag{9}$$

Again letting $\xi_k = \ln(R_k/r_k)/L$ and $\eta_k = (\theta_k - \sigma_k)/L$, then

$$\beta_k = \{(\xi_k + \eta_k) - i(\xi_k - \eta_k)\}/2L \tag{10}$$

or $\beta_k = \mathrm{Re}(\beta_k) - \mathrm{Im}(\beta_k)$

However, β_k should be real. Furthermore,

$$\beta_k = \sqrt{k} \cdot \beta_1 \tag{11}$$

Hence hydraulic diffusivity is given by:

$$\alpha = \frac{\omega}{2\,\{\mathrm{Re}(\beta_1)\}^2} \tag{12}$$

Given the viscosity and compressibility of fluid,

$$K/\phi = \frac{\mu C\,\{\mathrm{Re}(\beta_1)\}^2 \cdot T}{\pi} \tag{13}$$

Applications of the Model

As mentioned earlier, this model is prepared for use in connection with the flow of single, homogeneous fluids in the porous medium. When two adjacent and communicating observation wells are available in waterfilled sands, the oscillating water-level readings are converted of bottom hole pressures. However for gases, the bottom-hole pressures must be converted to pseudo-pressures (real gas potentials) according to Equation (2).

Verification of Model

Two examples of practical importance are considered. The first, which is designed to simulate oscillations in the aquifer surrounding the gas storage pool, serves to check the accuracy of the calculated result in comparision to the value assumed in the analytical functions and also to examine the sensitivity of the calculated results to errors in the input data. The second case is a field example and consists of pertinent data taken from a gas storage reservoir.

Case 1

The following functions are assumed in the aquifer (Figure 2):

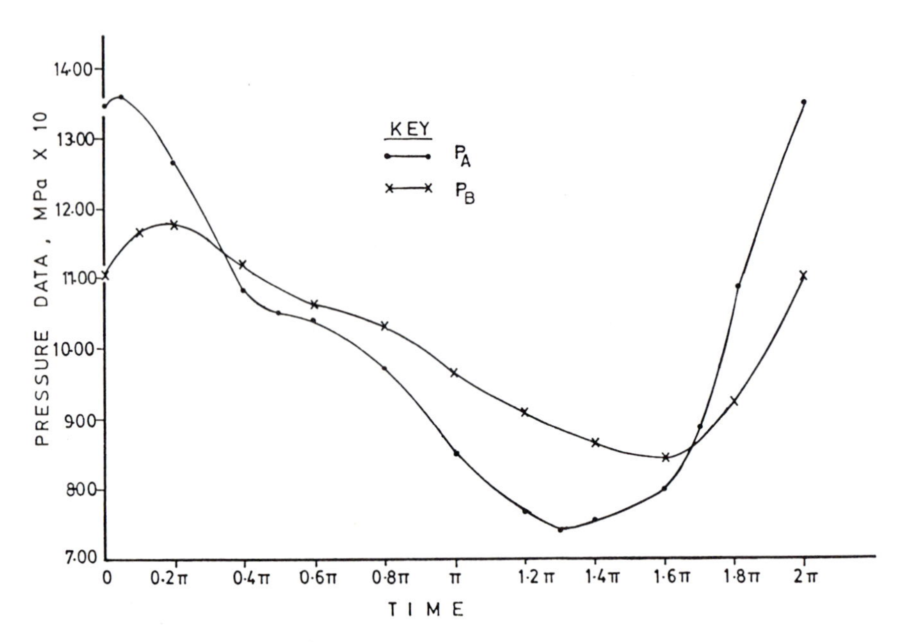

Figure 2. Pressure data for Case 1

Upstream Well A: $P_A = 1000 + 200\cos\omega t + 150\sin\omega t + 100\cos 2\omega t + 50\cos 3\omega t.$

Downstream Well B:

$$P = 1000. + 200\exp(-\beta_1 L)\cos(\omega t - \beta_1 L) + 150\exp(-\beta_1 L)\sin(\omega t - \beta_1 L) + 100\exp(-\beta_2 L)\cos(2\omega t - \beta_2 L) + 50\exp(-\beta_3 L)\cos(3\omega t - \beta_3 L)$$

Other pertinent data are:

Period of oscillation $= 2\pi$
Time interval taken $= 0.02\pi$, $\beta_1 L = \pi/6$ (0.5236)

The calculated values for $\beta_1 L$, $\beta_2 L$ and $\beta_3 L$ are shown on Table 1.

Table 1 Calculated Values of $\beta_k L$ (k = 1, 2, 3, ...)

k	$\beta_k L$
1	0.5307
2	0.7503
3	0.9309

The following are observed:

1. The calculated value of β_1 L is close to the assumed value ($\pi/6$) having a deviation of 1.36 %.

2. The calculated value of β_2 L is close to $\sqrt{2} \cdot \beta_1$ having a deviation of -1.33 %.

3. The percent deviation shown in 1. or 2. above represents the error introduced by the Fourier approximation scheme used on the data.

The possibility of errors in the field data is usually present. This is sometimes caused by faulty measuring devices and/or human judgement. To investigate the sensitivity of the calculated result to such errors in the field data, the upstream data (forcing function) is altered by 5 %, 15 %, 25 % and 35 % respectively. The deviations of the β_1 L values calculated with the skewed data from the β_1 L values calculated with the original data are determined and plotted on Figure 3 as a function of the skewness in the data. Figure 3 shows that the calculated result β_1 may alter very significantly when there are errors in the input data.

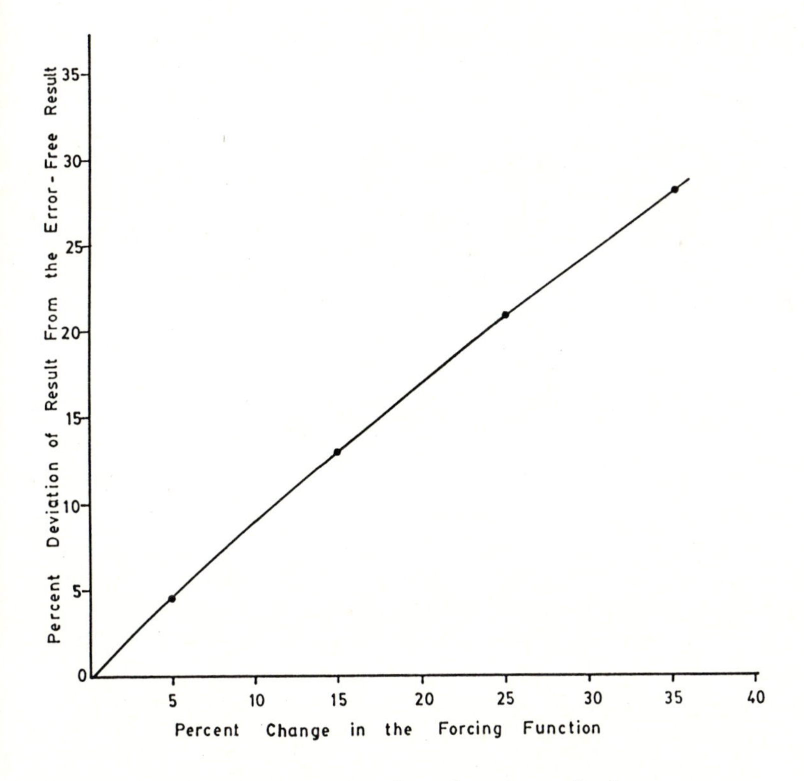

Figure 3. Sensitivity of calculated result to errors in the upstream well pressure data

Case 2

Bottomhole pressure data on the observation wells are shown in Figure 4. Other pertinent data are:

Distance between the two wells, x = 0.40 km (0.25 miles)
Relative density of gas = 0.6
Period of oscillation, T = 231 days
Datum depth = 960.00 m (3,150 feet).

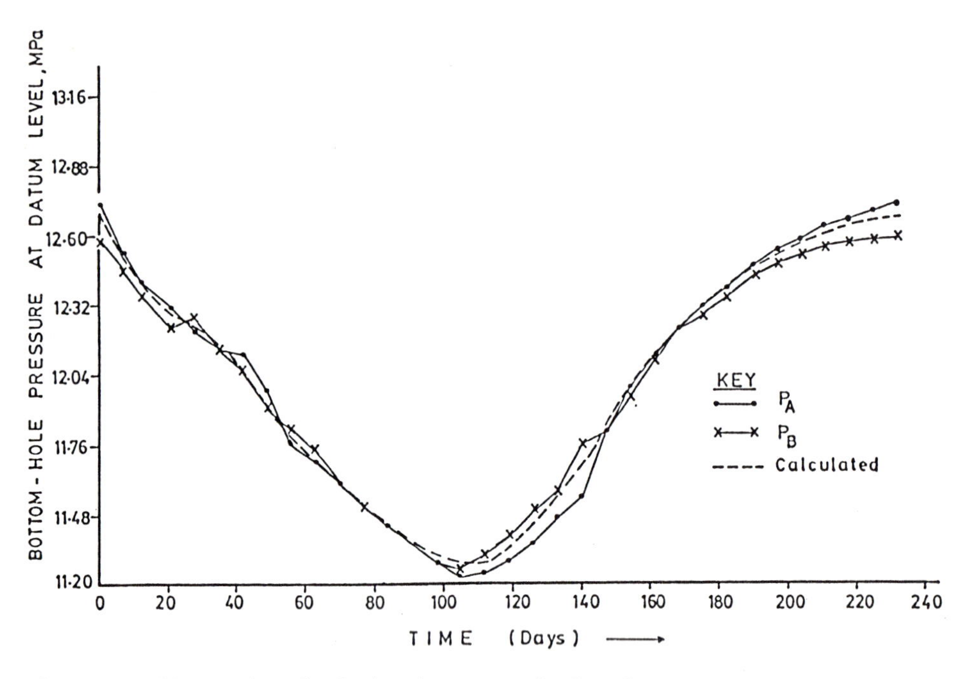

Figure 4. 'Observed' and calculated pressures in Case 2

Calculation of Hydraulic Diffusivity

Values of β_k from the computations are shown on Table 2 for k = 1,2, and 3. It is seen that $Re(\beta_1) = 0.1782 \times 10^{-5}\ cm^{-1}$. According to (12), hydraulic diffusity

$(\alpha) = \omega/2\ \{Re(\beta_1)\}^2 = \underline{49.56 \times 10^3}\ cm^2/sec.$

Taking C = 7.35×10^{-3} atm-^{1}
= 0.10
= 0.015 cp,
permeability (K) = 0.546 darcy (or 546 millidarcies)

This value is within range of probable values for formation permeabilities in aquifers and hydrocarbon reservoirs.

Table 2 Calculated Values of β_k (Case 2)

1	$.1782 \times 10^{-5}$	$.7267 \times 10^{-6}$
2	$.7292 \times 10^{-6}$	$-.5143 \times 10^{-5}$
3	$.7608 \times 10^{-5}$	$-.1631 \times 10^{-4}$
4	$.1433 \times 10^{4}$	$.1106 \times 10^{-4}$
...	....	....

Reconstruction of Pressures at the Downstream Well Location

The reliability of the above result is verified by reconstructing the downstream well pressures with the just calculated value of hydraulic diffusivity in the transient pressure equation (6) and checking how well the downstream well pressure data is matched.

Figure 4 also compares the calculated pressures (dotted curve) to 'observed' pressures at the downstream well (P_B). The average percent deviation is 0.062 and the standard deviation is 0.194 %. This agreement is remarkably encouraging and shows that the derived model may be used both in the determination of the effective reservoir hydraulic diffusivity from the sinusoidal pressure history and in the prediction of reservoir pressures at locations in the reservoir if effective formation properties are available.

CONCLUSION

1. The derived model permits the calculation of the effective reservoir hydraulic diffusivity and consequently permeability (K), or permeability/porosity ratio (K/ϕ) from sinusoidal pressure data on two adjacent and communicating wells in a reservoir. The present method does not require the cessation of normal reservoir operations.

2. Application of analytical functions in the model permitted comparison of the calculated result to the assumed value of $\beta_1 L$. The sensitivity of the calculated result to errors in the input data was also studied. The calculated result altered very significantly when errors were introduced in the input data.

3. Field application of the model in a natural gas reservoir gave a reasonable result which was later used in the model to reconstruct the pressures at the downstream well location. These reconstructed pressures showed an average deviation of 0.062 % and a standard deviation of 0.194 % from the given downstream well pressure data indicating very good and satisfactory agreement.

4. The present model may also be used in the prediction of pressures in the reservoir if the effective reservoir formation properties are available.

REFERENCES

Energy Resources Conservation Board, Alberta, Canada, 1979: Gas Well Testing: Theory and Practice, 4th ed.

Horner, D.R., 1951: Pressure Build-up in Wells Proceedings, 3rd World Petroleum Congress, Section II, 503-521.

Johnson, C.R., R.A. Greenkorn and E.G. Woods, 1966: Pulse-Testing: A New Method for Describing Reservoir, Flow Properties Between Wells. JPT, 1599-1603.

Kaplan, W., 1981: Advanced Mathematics for Engineers. Addison-Wesley Publishing Co., Reading, Massachusetts.

Katz, D.L., M.R. Tek, K. Coats et al., 1963: Movement of Underground Water in Contact wiht Natural Gas. AGA,New York.

Mueller, T. and P. Witherspoon, 1965: Pressure Interference Effects within Reservoirs and Aquifers, JPT, 471-474.

Udegbunam, E.O., 1983: Migration of Natural Gas from Storage Reservoirs. Ph. D.Thesis, University of Michigan.

Geological Appraisal of Groundwater Explotation in the Eastern Niger Delta, Nigeria

L. C. Amajor
Geology Department, University of Port Harcourt, P.M.B., Port Harcourt, Nigeria

Kewords: Niger Delta, Groundwater Exploitation, Borehole, Aquifers, Grain size analysis

ABSTRACT

This study shows that geologic considerations are rarely or inadequately incorporated into the design of boreholes, especially screen and pump characteristics, in the eastern Niger Delta hydrogeological province and, indeed, much of the country. Consequently, the geographic and stratigraphic distributions of wells can at best be described as very random. This has resulted in a fairly high rate of borehole failure, contamination and poor performance and may subsequently lead to more serious long term geologic hazards (salt water encroachment, subsidence and structure-failure and regional contamination by pollutants).

A certain degree of governmental intervention and control is advocated so as to achieve a good groundwater resource development and management plan for the country, protect customers and check the excesses of contractors whose primary interests are on financial gains rather than performing a good job.

INTRODUCTION

Uncontrolled increase in the rates of population growth, urban-drift, agricultural and industrial developments in the country have placed undue burden on our surface water resources. The Federal and State Governments, Industries, Institutions, communities and individuals have turned to groundwater for relief. This is buttressed by the revenue of N 1313 m (one thousand three hundred and thirteen million naira) allocated to boreholes by the Federal and State Governments for the 1981-85 plan period (Uzuchukwu, 1981). In the phase I borehole development programme each of the 19 States was given 40 wells, by the Federal Government. Thus, boreholing has become a booming business for professionals (Engineers and Geologists) and non-professionals alike. Consequently, many wells are cited and drilled anyhow and anywhere without due considerations to geology and the huge financial investment involved.

In the lower Niger Delta hydrogeological province, of which the study area is part (fig. 1), the main aquifer is the parali-continental Benin Formation. It consists of a porous and permeable succession of unconsolidated fine to coarse sand and gravel with minor amounts of clay. A thickness of 6000' (2000 m) is attained in places. Deposition occurred with the southward (seaward) progradation of the delta since mid Tertiary times, (Reyment, 1965).

Generally, groundwater is withdrawn from the upper (150 to 250 ft. (50 - 85 m)) section of the aquifer through 4 to 20 inches borehole size from over 1000 boreholes.

Based on consultancy experience with boreholes in the area, this paper discusses groundwater exploration and exploitation practices in the eastern delta (southern parts of Imo and Rivers States) from a Geologist's point of view and highlights the neglected geologic aspects of boreholing.

HYDROGEOLOGICAL INVESTIGATION

This pre-drilling or pre-construction engineering study utilizes indirect geophysical (resistivity, seismic), and photogeologic (Slide Looking Airborne Radar) methods to provide the following invaluable hydrogeological information:

a) location of potential aquifers and their recharge and discharge zones.

b) determination of their geometries (depth, lenght, width, thickness trend etc.).

c) expected lithologic sucessions and structure to be penetrated, and in coastal areas locates the fresh/salt water interface.

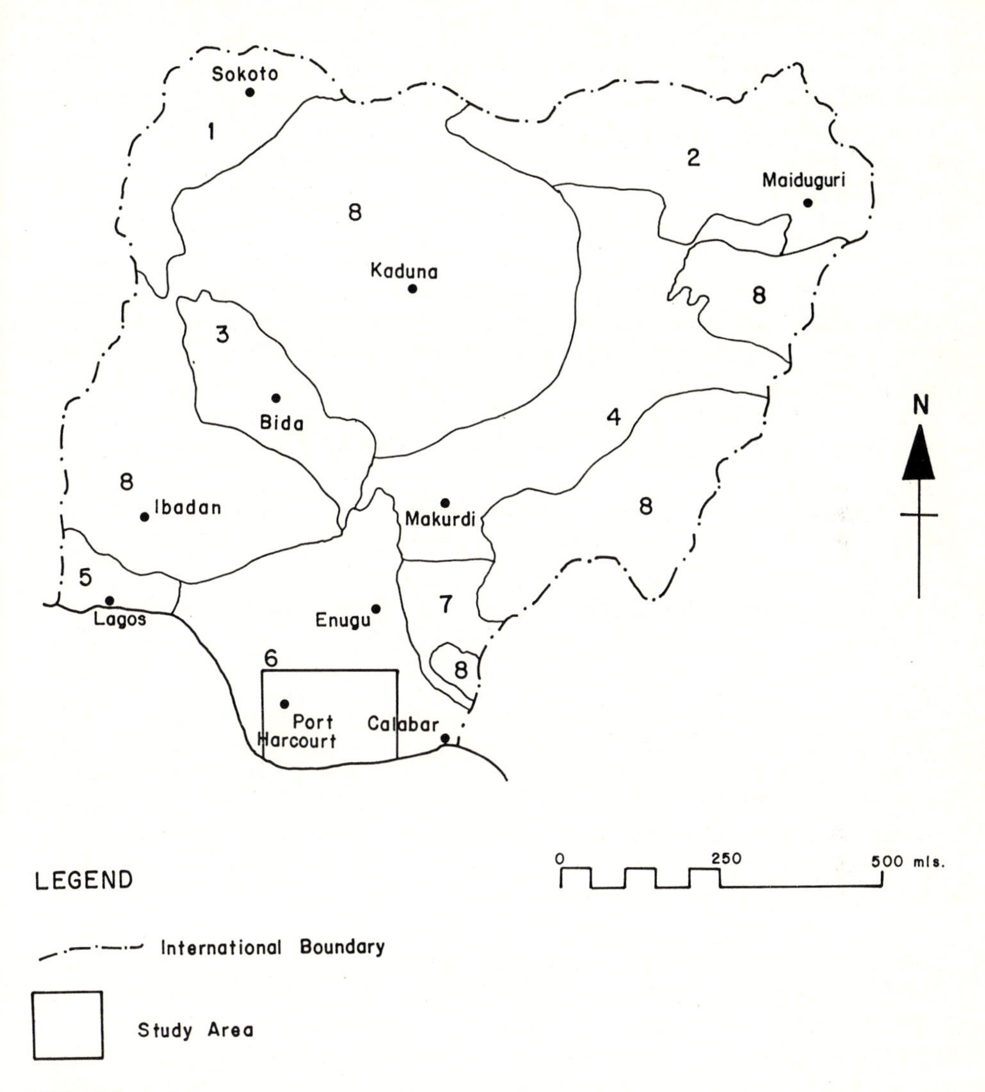

Figure 1.

Hydrogeological provinces of Nigeria
1-Northwestern (Sokoto sedimentary area), 2-Northeastern (Chad sedimentary area), 3-Niger (Upper Niger sedimentary area), 4-Benue (Benue sedimentary area), 5-Southwestern (Ogun/Oshun Sedimentary area), 6-Southcentral (Lower Niger sedimentary area), 7-Southeastern (Cross River sedimentary area), 8-Crystalline areas

d) potential pollution zones and water quality can also be inferred.

Information generated from test boreholes confirm or negate the findings of the above indirect methods as well as determine aquifer characteristics (porosity, permeability, water saturation, texture, continuity, transmissivity, depositional environments, subsurface lithologic profiles. quality of groundwater, etc.) Hydroglogical investigations determine recharge and discgarge areas as well as their rates (Adeniji, 1978).

This very important and necessary phase (groundwater exploration) preceeds any local or regional groundwater resource exploitation scheme. It provides a scientific base for good groundwater resource management as it provides pre-drilling information useful in borehole designs.

In the eastern Niger Delta, and indeed much of the country, this important preliminary phase to borehole construction is rarely carried out and where done scanty and almost useless information is available (e.g. see figure 2 and Table 1) (Balasha-Jalon, 1979; B.R.G.M., 1979; Intesca, 1977; Nedeco, 1961; Scan Water Engr., 1981).

In the study area there is the general assumption that the groundwater flow is from north of Port Harcourt (Coastal Plain Sands) towards the south and southeast to areas of discharge near the coast and swamps. This may not necessarily be correct for the whole region. This is because there is no regional water table map or flow net. Futhermore, the depositional environment is predominantly continental (Reyment, 1965) and this presupposes that most of the aquifers will be laterally restricted in time and space and young south-ward in the direction of delta growth. It is, therefore likely that the flow net will show varied flow directional patterns. The number and kinds of aquifers in the formation are not known and most importantly the fresh/saline water interface is not precisely known.

Most of the time, pre-drilling information, is not easy to come by because little co-ordination exists between the drilling companies and governmental agencies who normally award the hydrogeological investigation contracts. Most of the contracts are awarded to foreign companies or indigenous ones who turn to foreigners to do the jobs, and in most of them the outcome does not justify the financial reward. Most importantly, local experts are not given a hand at least to critically assess the reports before final payment. The net result is random distribution and drilling of boreholes which may lead ultimately to short and/or long term serious geological consequences.

BOREHOLING

Boreholing is the art of drilling a bore through rocks in the subsurface with the sole objective of producing potable water from water-bearing formations. The operation

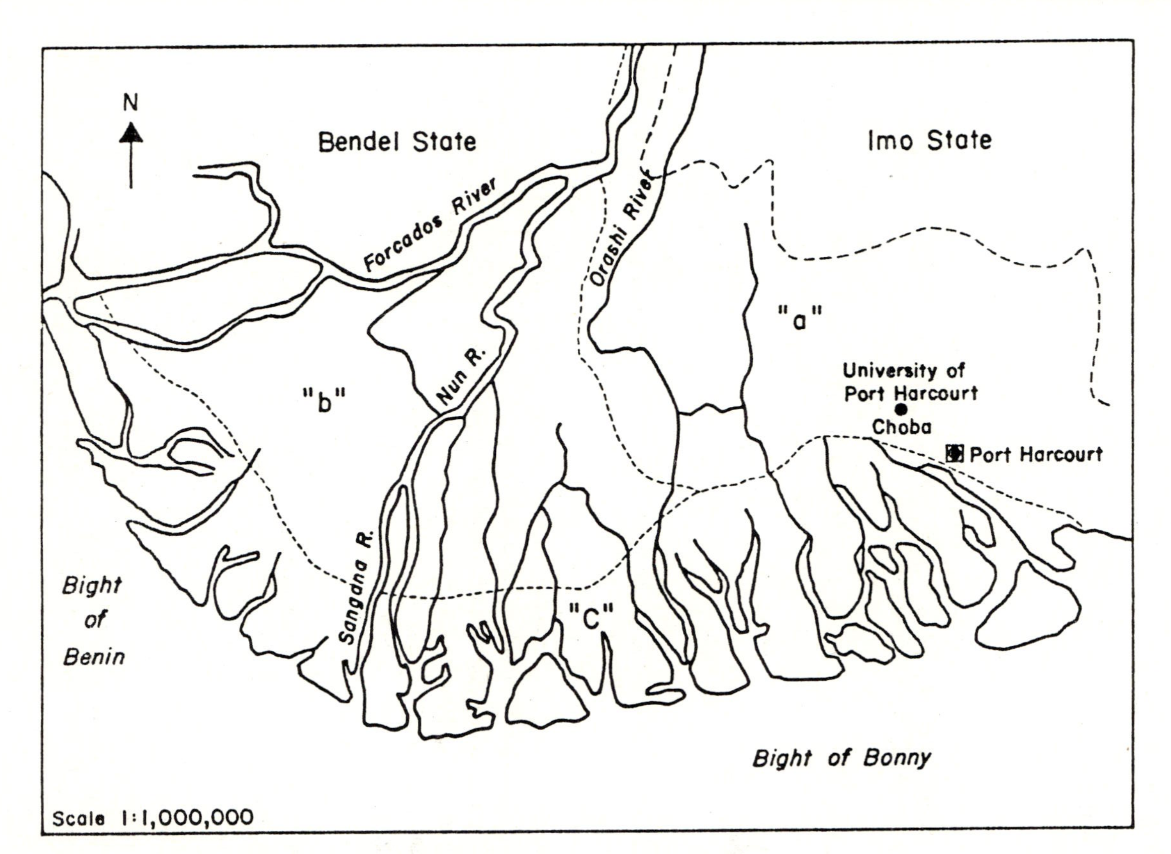

Figure 2. Niger Delta hydrogeological province
a-Shallow water aquifer, sand and gravel, good water, b-Shallow to deep aquifer, mixed sand and clay, high iron content water, c-Deep sand aquifer, high iron water.

Table 1 Results of hydrogeological investigation on the groundwater provinces of Nigeria.

Ser. No.	Hydrogeological Area	Aquifer	Discharge	Remarks
1.	Northwestern	Gundumi/ Illo	53,000	Widespread, slowness of recharge, moderate yields.
		Rima group	100,000	Freely recharged from outcrop and seasonal rivers
		Sokoto formation	48,000	Prolific, not extensive
		Gwandu formation	116,000	Freely recharge from outcrop and seasonal rivers
2.	Northwestern	Chad formation	-	Deep and probably not replenished
		Middle zone	-	Very extensive, found at moderate depth
		Upper zone	-	Good quality not corrosive, recharged locally
		Sands of the Bima ridge	-	Narrow, shallow, limited storage and recharged annually.
3.	Niger	Sandstone	-	Underlie the entire hydrological area, moderate depths
4.	Benue	-	-	-
4-1	Post Cretaceous sediments	Kerri, Kerri formation	-	Miderate depth high potential
4-2	Upper Benue	Lower Cretaceous sandstones	-	Low storage, extensive jointing recharge through percolation
5	Southwest	Abeokuta fm.	-	Extensive, high storage and transmi-ssivity from content high in deeper state.
		Coastal Plain Sands	-	Has local salinity problem along the axis of Ogun fault
6	South-Central			Abundant water of good quality. Recharge not quantified yet. High potential
6-1	Cretaceous sediments			
6-2	Post-Cretaceous	Sediments	-	Underlain by thick, permeable, s-rongly recharged aquifers
7.	Southeastern			Small amount of groundwater available from joints. Area not regarded as a major groundwater area.
8.	Crystalline Area	-	-	Not regarded as groundwater develop-ment areas. Preferred for major supplies, surface impoundment and abstraction.

is normally carried out in three main stages: Construction/drilling, Completion and Rehabilitation.

Construction: Cable tool and hydraulic percusion, mechanical and hydraulic rotary (normal and reverse circulation), and combined percussion-rotary, are some of the drilling methods used in groundwater exploration. The choice of a particular method depends on:

a) the purpose of the well

b) character of the rocks to be penetrated

c) and the quality of water required, depth to aquifer and, of course, cost factors.

Because the sands of the Benin Formation are unconsolidated, the conventional rotary method, which is faster and economical, is widely used. Mobile truck mounted hydraulic and mechanical (tripod) rigs are used in dry upland (upper delta) and swampy (lower delta) areas, respectively, because of the inherent transportation problems dictated by the terrane. This is good drilling practice. However, this technique requires a good choice of drill bit, stem and fluid and this is dicitated by the geologic conditions to be encountered in the area. Because most of the time this knowledge is lacking incidents of hole collapse, stuck drilling pipes, well development problems, poor sample recovery and unusual longer drilling time (up to 6 months) are very common in the study area, especially with the tripod method of drilling. Most commonly contractors get away with these by telling all kinds of "cock and bull" stories to their unknowledgeable clients like a thick bed of stone was encountered. In some cases this hypothesis enables them to ask for more money in preparation to leave where the client fails to honour their demand especially in purposely undercosted wells.

Usually as drilling is in progress samples from the formations penetrated especially the aquifer, are recovered. These are adequately described lithologically, minerallogically and texturally in strict compliance with geologic terminology and nomenclature. This is later transformed into a pictorial form known as a lithologic or drillers log, which shows the variations of the above parameters with depth. Electric well logs (caliper, SP and Rt) should be run in uncased holes. These also provide varied useful information on some aquifer characteristics (Smith, 1977). Integration of all these pieces of information lead to the establishment of the continuity of the aquifer, subsurface geologic profiles, hydraulic characteristics and depositional environment of the aquifer, water quality and are useful in subsequent drilling stages and future maintenance of the well.

In the study area the above procedure is hardly ever followed. In a few cases only verbal lithologic and textural descriptions which make little geologic sense is carried out and rarely are electric logs run. However, when they do, measurements are normally made at intervals of 5 to 10 feet. Thus, the electric log is not a continuous recording of the variation of the parameters with depth, because lots of interpolation are inherent in the method.

Well Completion

After constructing or drilling the borehole, it has to be completed. This involves screening the intake section of the aquifer, casing the remaining portion of the well up to a few feet above ground level and well development.If these are properly designed they make for excellent yield and long life.

Screening

The most important screen characteristics are length, diameter, strength and slot size.Choice of the right combination of screen parameters for any well is dictated by water quality, aquifer thickness and grain size characteristics. Screen choice, therefore, is dependent on aquifer geology. Thus, there are hydrogeological rules that govern their selection based mostly on granulometric analysis of aquifer samples. For example it is good practice to screen the bottom 1/3 and 70 to 80 % of the aquifer thickness in a homogenous water table and artesian aquifer, respectively. Futhermore, the screen slot size for a homogenous fine sand aquifer is that size that retains 40 to 50% of the sand. In general, screened boreholes perform best when the total intake area of the screen is as great as possible for a given slot size and strength requirement (Johnson, 1975).

In the study area screen length is determined before drilling. The common practice is to install 10 to 20 feet of screen length without regard to aquifer thickness and purpose of the well (Ugboaja, 1983). The size of slot opening on the other hand is rarely determined through the results of grain size analysis (Okoli, 1983). These are done for economic considerations without considering seriously the groundwater situation or the clients present and future well requirements. This, is to me, more of a dis-service to the owner than to over design. Thus, many wells have been abandoned because they pump sand and water, are plogged resulting in unexpected low yield and shattered pumps in the hole due to emergence.

Casing

These are pipes of various sizes and composition used to protect the unscreened portion of the well from collapse. It serves as a housing for the pump and a conduit for water flow upwards to the surface. Casing size is dictated by the bowl size of the pump and expected water yield and water quality. Water contamination can occur through corrosion of any part of the casing (WHO, 1978). For instance, a pump expected to produce 150-400 gpm should have a minimum and maximum casing size of 8" ID, respectively (Johnson, 1975).

In the eastern delta, API steel, inferior galvanized ("Thaiwan") steel and plastic pipes have been used in places. However, the choice of casing diameter is never based on the expected yield. The use of cheap inferior casing materials has contaminated lots of boreholes. This, in turn, has given rise to many pumps being lost in the hole (i.e. fall to the bottom of the well and cannot be fished out). The common pracitce is to install the inferior casing pipes when the well owner is not around.

Gravel Packing

This is the introduction of artificially graded siliceous gravels around the screen to make the area more permeable and thereby increase the effective diameter of the well. This excercise is conducted for thick artesian, unconsolidated, uniform fine sand and laminated aquifers. The grading of the gravel is selected on the basis of the results of grain size distribution study on the aquifer samples. The size of the screen slot is then controlled by the size of the gravel (Johnson, 1975).

Gravel packing is commonly practiced in the study area. However, the size of the gravel and screen slot width are not based on the results of mechanical analysis of aquifer and gravel samples. Such a practice has caused lots of development problems for many water drilling contractors in the area.

Grouting

In a well designed well, cement grout (mixture of cement and water only in the ratio of about 800 litres per m^3 of cement) not less than 8 cm in thickness is used to fill the annular space between casing and the wall of the drilled hole for the entire casing length. This ensures the stability of the borehole and more importantly, prevents borehole contamination from outside the intake (screened portion) area. Geologic conditions, therefore, determine depth of grouting.

The unusual grouting practice of most water drillers is to fill the annular space with cement bags and cement grout from 2 to 10 feet below the surface. In some cases sand is added to the grouting mixture. Thus, most boreholes in the area cannot be said to be contamination proof as the water table is shallow 0 - 15 m (Ala, 1981) and the general environment very hazardous with lots of pollution sinks from industrial and oil company activities.

Development

This process utilizes mechanical surging, or swab, or over-pumping or compressed air or high velocity water jetting methods to declogg the aquifer of drilling fluid and finer sediments. The net result is increased porosity, permeability and stability of the sand around the screen so that the well produces sand-free-water at the maximum capacity (Johnson, 1975). A sucessful development program depends more on a better screen design and therefore geologically controlled.

Compressed air is commonly used for borehole development in the study area. In a few cases jetting with water at high velocity has been sucessful. Both methods are acceptable but the latter is more effective, and advantageous. Because of the poor screen design discussed earlier, development has been unsucessful in many boreholes. In some, water and sand are continuously pumped out, in others the aquifer yield is unexpectedly very poor, and at times, development time is unduely too long.

Pump Selection and Installation

A preliminary pumping test on the borehole determines yield and draw down characteristics of the aquifer, and provides the only sound basis for the selection and purchase of the elements of a permanent pump and pumpage suitable to the operating characteristics of the well. This reduces pump and pumping costs, maximizes satisfactory pump performance and extends the life of the well (Heindle, 1975).

Pump test is rarely conducted especially, in private boreholes. Where attempted, the results are very unreliable for any meaningful predictions to be based on them (Ugboaja, 1983). It is common practice to install 3 to 7.5. HP pumps in wells on no basis. It is therefore not suprising to hear clients complain that water cannot be pumped into their overhead tanks and that water stops to flow out from their wells after some pumping period. Investigations on these complaints reveal that in the first case a low HP pump was installed. In the second case the depth at which the pump was installed was such that the pump was emergent after some time. This has led to the destruction of many pumps.

Well Rehabilitation

After a borehole has been completed it is good practice to wash it with clean water or chlorinated water or treat it with polyphosphates to ensure that potable water is produced. Water quality analysis determines whether treatment is necessary and the treatment method to employ.

Table 2: Water quality parameters commonly analysed for in study area compared with those recommended by WHO (1971).

Parameters	WHO drinking water Std.	Study Area
Total Solids	"	"
Color	"	"
Odour	"	"
Turbidity	"	Total dissolved solids
Chloride	"	Chloride as Cl & Nacl
Iron	"	"
Manganese	"	"
Copper	"	
Zinc	"	
Calcium	"	
Magnesium	"	
Sulfate	"	"
Total hardness (as $CaCo_3$)	"	"
Nitrate (as No_3)	"	
Phenol	"	
Anionic detergent	"	"
Fluoride	"	"
pH	"	"
Arsenic	"	
Cadmium	"	
Chromium (6+)	"	
Cyanide	"	
Lead	"	
Mercury	"	
Selenium	"	
Polynuclear aromatic hydrocarbon	"	
Alpha radioactivity	"	
Beta radioactivity	"	
Coliform bacteria	"	

Well rehabilitation is hardly ever practiced in the study area. As soon as the borehole is completed, and pump installed, the borehole is capped and the contract is deemed completed. In most cases, water quality is not determined and where conducted, the analyses is never complete Table 2.

Generally though, most wells in the area produce iron-rich water (Etu-Efeotor, 1981). Whether the source of iron is due to corrosion of the pipe and/or screen or to aquifer contamination remains to be proved. To eliminate the former it may be necessary to design boreholes with fibre glass casing and stainless steel screen (Decker and Russel, 1977).

AN EXAMPLE OF A CASE STUDY

The University of Port Harcourt boreholes will be used to illustrate some of the issues discussed in the text. Seven boreholes were drilled in the campus between 1977 - 1982 to supply groundwater to the University community. Only five are currently productive and each yields about 10,000 gallons per hour. Table 3 shows the design characteristics of some of the boreholes. Groundwater is produced from the upper secton of the Benin Formation between 210 and 390 feet depth through 7" boreholes cased with API steel. The writer believes that the following design errors were made:

1. The length of all the screens used is 20 ft and aquifer thickness is not recorded. My investigation revealed that aquifer VI is 150 ft thick and a 50' long screen set at the base of the aquifer should have been a better design.Futhermore, well IV should have been screened between 280-300 ft depth.

2. Aquifer grain size analysis was carried out only for wells II and VI and in each case, only one sample from a particular horizon in the aquifer was analysed. This sample, is therefore, not representative of the aquifer. All samples recovered from the aquifer should have been mixed and analysed or samples from about 5 ft interval analysed and results integrated to arrive at the best slot size.

3. Screen slot size for four of the wells is 0.2 mm and 0.04 mm for the fifth. Calculations based on analysis of the aquifer samples from wells II and IV show that slot sizes of 0.8 and 0.9, respectively, should have been better and 0.5 mm for well VII. Absence of grain size data for the other boreholes suggests indiscriminate choice of screen slot size.

4. Pump test was conducted for wells II and IV only suggesting the indiscriminate choice of pump parameters for other boreholes. My estimations indicate that pump setting depths for wells II and IV should have been 110 and 120 ft. respectively. In the latter well (IV) the depth interval 98-125 ft. was screened and pump set at 180 ft far below the screen depth. Besides, the pump powers (10 and

Table 3 Design characteristics of University of Port Harcourt boreholes.

Borehole	Drilling Contractor	Total depth (ft)	Casing type	Casing size (in)	Screen length (ft)	Screen size (in)	Slot size (mm)	Depth screen set (ft)	Discharge (gph)	Draw-down (ft)	Specific capacity gph/ft.	Total dynamic head (ft)	Pump power HP	Bowl diam pump (in)	Setting depth of pump (ft
I	West African Boreholes (NIG.)Ltd.	210	API Steel	7	20	6 5/8	0.2	117-137	10,000	39.27	254.65				
II	Water Pit (NIG.)Ltd.	260	"	7	20	-	0.2	240-260	8,000	-	-	26.5	10	3	160
III															
IV	Wedcon (NIG.)Ltd.	300	"	7	20		0.2	98.5-125	10,200	15	680	28.5	12	3	180
V															
VI	Water pit (NIG.)Ltd.	390	"	7	20	6 5/8	0.2	260-280	10,000	122	81.97	-	-	-	-
VII	Water pit (NIG.)Ltd.	220	"	7	20	6 5/8	0.04	199-219	10,000	-	-	-	-	-	-

12 HP) are considered too high for the total dynamic head (26.5 and 28.5 ft.). This to the writer, is overdesign.

5. No data on wells III and V was submitted by the contractors to the client. The boreholes are located in the shallow water aquifer zone 'a' of figure 2. This zone is expected to yield good water, however the wells produce iron-rich water (Ugboaja, 1983; Okoli, 1983). This suggests that the pre-drilling investigation results are not too correct. This, however, is not without prejudice to the alternative of contamination by corrosion of the pipes.

6. Boreholes I and III have failed and are currently not producing. Attempts to fetch out the pumps have also failed. Tha cause is not known. Borehole collapse, screen clogging, lowering of water table below depth pump was set and extensive pipe corrosion are suggested as possible causes of failure of the boreholes.

7. Casing and pump sizes are, repectively, 7 and 6 5/8 inches. For a yield of about 175 gpm 6 inch pump and 8-10 inch internal diameter casing should have been used.

CONCLUSION

1. The design of most boreholes, especially their screen and pump characteristics, in the eastern Niger Delta hydrogeological province is very poor because geological conditions are not taken into consideration. Most importantly, pre-drilling hydrogeological investigations are lacking but where conducted the results are unreliable and not readily available to Water Drilling Companies. Consequently, the problems of well failure (lost pumps, collapsed wells, stuck pipes, etc.), high costs of pump and pumping, contamination and unexpected low water yield are bitterly borne by the customer.

2. The indiscriminate sitting of boreholes to tap the upper part of the Benin Formation could lead to some serious geologic problems in the long run. These include salt-water intrusion, subsidence and possible failure of structures, contamination of aquifers by pollutants and dry wells due to well interference, overpumping and discharge exceeding recharge.

3. It is recommened that governments should step in and control the borehole business to some extent, especially as any resource underground belongs to the Federal Government and to protect clients from "money -hungry", water borehole contractors. A governmental agency will beside other things perform the following duties.

a) Ensure that all registered water well companies have at least a construction engineer, a geologist and an experienced driller,

b) Determine the hydrogeological conditions of any area because past development of most aquifers has been in ignorance of their potentialities and especially their limitations,

c) This will enable them determine and execute minimum acceptable distance between wells and other structures which, hitherto, is baseless and indiscriminate,

d) Supervise borehole construction and completion activities by contractors.

This will not only ensure good borehole design and maintainance but will facilitate the collection and deposition of hydrogeological data on each borehole in a central data bank. Such data, which should be made easily available to any person, will be used to improve upon earlier hydrogeological interpretation and predictions of an area made on the basis of pre-drilling studies. The ultimate result will be improved groundwater resource management,

e) determine and adequately manage the recharge areas of aquifers because to maintain groundwater resources indefinitely, recharge must balance or exceed discharge. Failure to pattern the development and management of groundwater in accordance with the availability of groundwater may result in overdeveloped or underdeveloped aquifers.

Finally, groundwater planning and management should be under the direction of knowledgeable and competent experts familiar with all the elements that make up good integrated and efficient groundwater resource plan and management scheme. Government intervention and control of the borehole business could be a potential source of revenue for the country. For instance, anybody who requires a private borehole should obtain a licence or permit on payment of some amount. Supervision of borehole construction and completion should be paid for. Pre-drilling investigation reports and other data should be paid for by contractors, researchers and consultants.

Where it is absolutely necessary to award groundwater contracts to foreign companies local experts should be involved in these projects and should critically evaluate their reports before final payments are made.

REFERENCES

Uzuchukwu, B.N.C., 1981: Paper presented at the International Conference and exhibition on soil investigation and groundwater for developing nations, Kuala Lumpur, Malaysia.

Reyment, R.A., 1965: Aspects of the Geology of Nigeria: Univ. Ibadan Press, Nigeria, 133.

Adeniji, F.A., 1978: Role of hydrology in drought prevention: Inaugural lecture of the Nigerian Hydrological Technical Committee, Maiduguri, Nigeria.

Eakin, T.E., D. Prince and J.R. Harrill, 1977: Developing groundwater resources: The Johnson Drillers Journal, Sept. - Oct., 1977,4-7.

Balasha-Jalon consultants (Nig.) Ltd., Benin, 1979: Hydrogeological network in the Niger delta area.

B.R.G.M. (Nig.) Ltd., Oyo State, 1979: Hydrogeological investigations in the sedimentary basins (Coastal Plains and Niger Delta, Vol. 1).

Intecsa, 1977: Proposals for pre-drilling hydrogeological investigations in Rivers State.

Nedeco, 1961: The waters of the Niger Delta.

Scan Water Engineering, Denmark, 1981: Proposal for new waterworks at Port Harcourt Township.

Smith, Ali.,1977: Nebraska studies ways to recharge aquifer storage: The Johnson Drillers Journal, Sept. - Oct. 1977, 1-3.

Johnson, E.E., 1975: Groundwater and wells: Johnson Division UOP Inc., Minnesota, 440.

Ugboaja, R.N., 1983: Groundwater survey around University of Port Harcourt: Unpubl. B.Sc. thesis, University of Port Harcourt, 63.

Okoli, M., 1983: Groundwater exploration and exploitation in the University of Port Harcourt: Unpub. B. Sc. thesis University of Port Harcourt, 51.

WHO, 1978: Water quality surveys: 352.

Ala, N., 1981: Hydrogeology of Port Harcourt and environs: Unpubl: B. Sc. thesis, University of Port Harcourt, 65.

Heindle, L.A., 1975: Hidden waters in arid lands: Intern. Development, Research Centre, 19.

Ete-Efeotor, J.o., 1981: Preliminary Hydrogeochemical investigations of subsurface waters in parts of the Niger Delta: Journal Min. Geosci. Soc., 18, 103-10.

Decker, T.L. and E.D. Russel, 1977: New design gives Denver district iron-free well: The Johnson Drillers Journal, Jan - Feb., 1977, 5-8.

Iron Deposits of Nigeria

M. A. Olade[1] and J. A. Adekoya[2]

[1]School of Earth and Mineral Sciences, Federal University of Technology, Akure, Nigeria
[2]Department of Environmental Sciences, Polytechnic, Ibadan, Nigeria

Keywords: Nigeria, Iron Ores, Ferruginous quartzites, Chemical analysis, Industrial development

ABSTRACT

Iron deposits of Nigeria occur within both the Precambrian basement complex and the younger sedimentary basins, in the form of ferruginous quartzites, banded iron formations and oolitic ironstones.

The ferruginous quartzites (Itakpe type) are metamorphosed iron-rich sediments that occur as bands and lenses within Precambrian gneisses and migmatites. The Itakpe deposit which contains more than 300 million tons of iron ore, averaging 40 % Fe will form the major source of raw material for the iron and steel plants.

Banded iron formation (Birin Gwari type) associated with low-grade metasediments, and metavolcanics of the Proterozic 'schist belts' occur in small units of considerable extent in northwestern and central Nigeria. Their economic potential are yet to be fully assessed.

Large reserves of minette-type oolitic-pisolitic ironstones often overlain by laterite ores have been developed in the Maestrichtian and Paleocene marine sequence of the Niger and Sokoto basins. Although, the proven reserves are large, the relatively high phosphorous and alumina contents will make special beneficiation necessary prior to utilization.

Nigeria's iron-ore deposits have considerable potential and can support a viable iron and steel industry.

INTRODUCTION

Iron and steel have often been regarded as the backbone of human civilization and the basis of industrialization. Also, the total production and per capita consumption of iron and steel products have been widely used as measures of the level of industrial development and living standards of nations (Park, 1974). Thus, for many developing countries like Nigeria, that are striving for rapid economic development, the possession of a domestic iron and steel industry has become a matter of crucial importance. In pursuit of this aspiration, two intergrated iron and steel mills (Aladja and Ajaokuta), and three rolling mills (Katsina, Jos and Oshogbo) have been established recently. When fully operational, these steel plants will require about 5 million tons of iron ore annually. Until recently, the iron-ore resources of Nigeria have not been seriously investigated, and so far the steel plants utilize raw materials imported from Brazil and Liberia.

The objective of this paper is to examine the geological environment and characteristics of iron ore deposits in Nigeria, with emphasis on their potential utilization as raw materials.

SYNOPSIS OF NIGERIAN GEOLOGY

Nigeria forms part of the reactivated Precambrian basement of the Dahomeyan sheild lying between the West African and Congo cratons. About half of the country is underlain by crystalline rocks comprising Archean gneisses, Proterozoic metasediments and Pan-African (Late Precambrian) granitoids. The rest of the country is covered by a veneer of Cretaceous - Recent sediments confined to well-defined intracratonic basins (Fig. 1).

The Archean gneiss-migmatite complex are the most wide-spread rocks within the rejuvenated crystalline basement. They are composed predominantly of quartzo-feldspathic biotite gneisses and migmatites with intercalations of para-amphibolites and quartzites, some of which are ferruginous. The Proterozoic metasediments and metavolcanics occur as well defined schist belts steeply infolded into the reactivated basement. The dominant lithologies are amphibolites, ultramafic schist, pelitic and semi-pelitic schists and ferruginous quartzites.

Both the basement and supracrustal cover rocks are, in places, intruded by a suite of Pan-African orogenic (500 m.y) granitoids known as the 'Older Granites' to differentiate them from the much younger and anorogenic 'Younger Granites' of Jurassic age confined to the Jos Plateau in Central Nigeria. Less common igneous rock types emplaced into the basement complex include charnockites, syenites, gabbros and serpentinites.

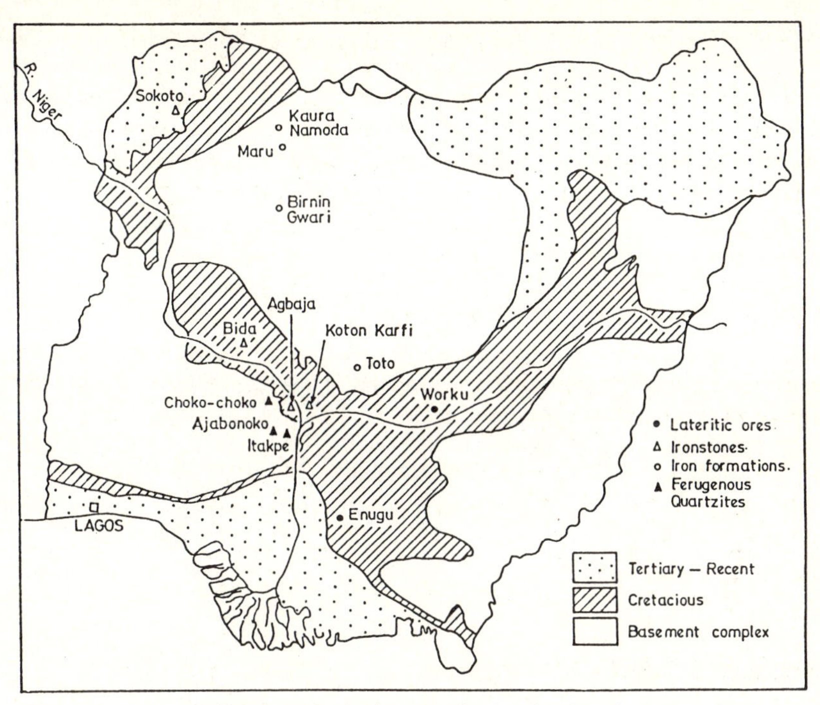

Figure 1. Generalized geological map of Nigeria showing the locations of various types of iron deposits.

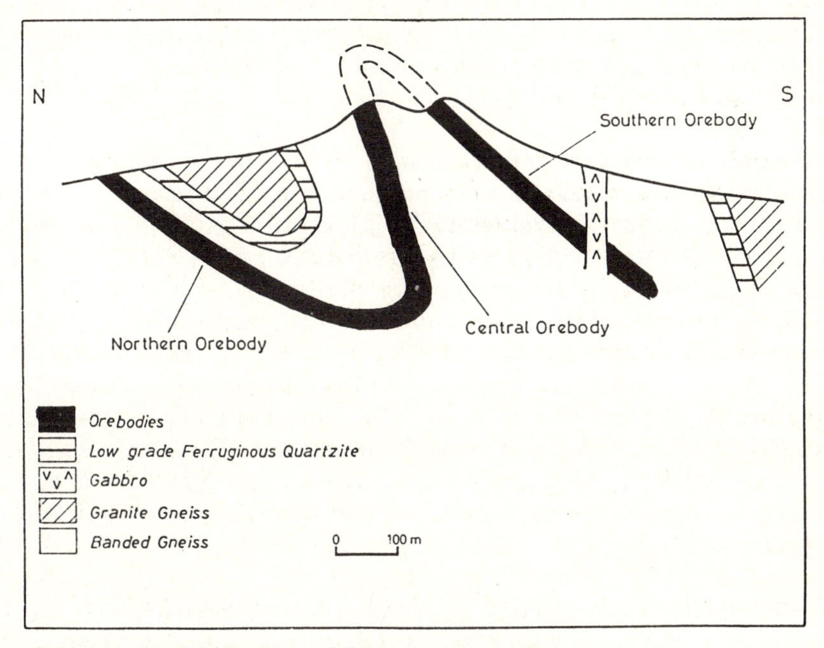

Figure 2. Geological map of Itakpe iron deposit.

Cretaceous sediments comprising marine shales, limestones and ferrigenous sandstones were the earliest sediments to be deposited within the Benue, Sokoto and Niger intracratonic basins. Of particular interest are the oolitic ironstones and ferruginous sandstones of Maestrichtian age located in the three cosanguinous basins. Tretiary to Recent sediments, mostly sandstones and shales, are confined to the Niger Delta and Dahomeyan Basin in the south, and the Chad and Sokoto Basins in the north.

IRON-ORE TYPES AND OCCURRENCES

Iron-ore resources of Nigeria can be classified into four main catagories based on geological setting and mode of origin.

a) Ferruginous Quartzites
b) Iron Formations
c) Sedimentary Ironstones
d) Laterites and Lateritic Ironstones

a) Ferruginous Quartzites

The ferruginous quartzites are metamorphosed iron-rich sediments consisting predominantly of iron oxides and quartz gangue that occur as bands and lenses within Precambrian gniesses and migmatites. The major deposits belonging to this category are those of importance as potential raw material sources for the Nigerian steel plants; Itakpe, Ajabonoko and Cho-Choko, all located around Okene-Lokoja district in central Nigeria (Fig. 1).

The Itakpe deposit which contains more than 300 million tons of iron ore occurs as steeply-dipping bands and lenses intercalated within gneisses and quartzites. Out of the three orebodies so far delineated, the central orebody is the most promising (Fig. 2). It consists of two relatively high grade ore lenses 35 m and 15 m thick and extending 200 m along strike. The northern orebody which occurs at the northern flank of the deposit thins from 60 m in the west to 25 m in the east. Petrographic studies (Olade, 1978) show that the ores are of three types, massive magnetite ore (4 % of reserves), banded or granular hematite-magnetite ore (80 % of reserves; Fig. 3) and schistose hematite ore (15 %). Ore grade averages 40 % Fe with values of up to 62 % Fe in the massive ore. Pilot metallurgical tests have shown that the ores can be easily beneficiated by magnetic and gravity methods to yield a concentrate with 65 % Fe. Pilot mining involving the development of several benches has commenced.

The Ajabonoko deposit is located just a few kilometres northwest of the Itakpe deposit (Fig. 1). It occurs as a series of highly fractured and gently-dipping pods and

Figure 3. Photograph of granular specularite-magnetic ore, Itakpe deposit.

lenses of ferruginous quartzites about 15 - 60 m thick and intercalated within gneisses and migmatites. The principal ore minerals are magnetite and hematite which form either thin bands alternating with quartz, or as homogeneous ore. Ore grade ranges from 42 - 56 % Fe with an average of 45 % Fe. Ore reserves are estimated at 60 million tons.

The Choko-Choko deposit lies about 35 km west of Lokoja (Fig. 1) and comprises several thin bands (2 - 5 m) within schists and quartzites that are intruded by granites. The principal ore mineral is magnetite (as a result of contact metamorphism) in a gangue of quartz. Average ore grade is 38 % Fe with estimated reserves of 15 million tons.

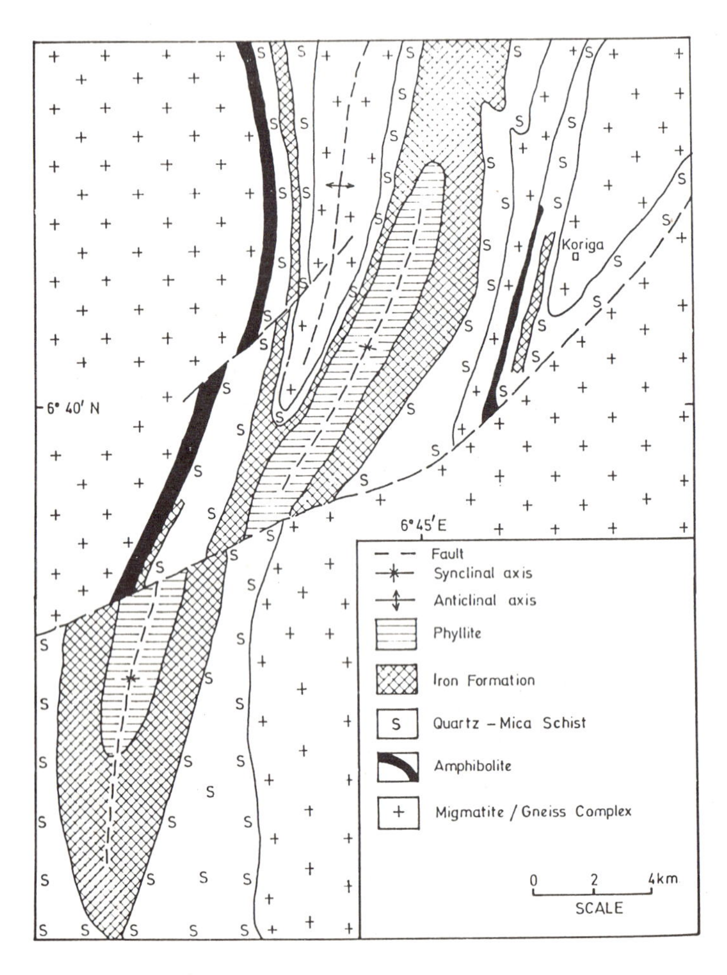

Figure 4. Geological map of Birnin-Gwari banded iron formations.

b) Iron-Formation

Low-grade banded iron-formations occur at several localities within the Precambrian 'schist belts' of northwestern and central Nigeria (Fig. 1). Notable among these deposits are those at Birnin Gwari, Maru-Kaura Namoda, and the recently discovered deposit at Tot in central Nigeria (Fig. 1). These iron-formations are products of low-grade metamorphism which commonly occur as narrow folded bands and lenses interbedded with pelitic and semi-pelitic schists, phyllites, metavolcanic rocks (Fig. 4; Baer, 1982; Adekoya, 1981).

The iron formations are predominantly fine-grained oxide facies, and are characterized by a rhythmic alteration of quartz-rich and iron oxide-rich bands or layers which vary in thickness from a few millimetres to 5 cm (Fig. 5). The very fine grained texture of the ores coupled with the vitreous appearance of the quartz suggests that the silica is metamorphosed chert (Adekoya, 1981). Hematite is the dominant ore mineral while magnetite (commonly martitized) occurs in subordinate amounts.

Although the iron-formation may contain large reserves of ore, the degree of mineralization varies considerably thereby producing rather low quality ores with grades ranging from 21 to 40 % Fe. No supergene enrichment has developed over the deposits compared to those in Brazil.

c) Oolitic ironstones

Oolitic ironstones of the Minette or Lorraine type are developed in the Upper Cretaceous sediments of the Niger and Sokoto embayments (Fig. 1). They represent a shallow marine transgressive facies in the sedimentary sequence. In the southern (Lokoja area) and middle (Bida area) parts of the Niger embayment, they occur in two stratigraphic horizons within a predominantly fluviatile - deltaic succession of Maestrichtian to Campanian age (Jones, 1958; Adeleye and Dessauvagie, 1972).

The upper ironstone froms a capping over the argillaceous sequence of Patti and Enagi Formations of the Lokoja and Bida areas respectively. Its thickness varies from 5 m near Bida to 15 m at the Agbaja plateau (Adeleye and Dessauvagie, 1972; Adeleye, 1976). The lower ironstone beds have a maximum thickness of about 5 m.

A continuous bed of the upper ironstone caps the Agbaja plateau and other mesas around Lokoja - Koton Karifi district covering a combined area of over 256 sq. km. The north-eastern part of the plateau was investigated by drilling. Ore reserves of 41 million tons assaying between 45 and 50 % Fe was delineated in an area of 1.46 sq. km. There is an overburden of laterite derived from the weathering of the ironstone which also constitutes a low grade iron ore having a variable tenor of 28.2 - 48.7 % Fe and thickness varying from 0.8 - 4 m (Fig. 6). In an adjacent locality

Figure 5. Photograph of banded iron formation showing deformed laminations.

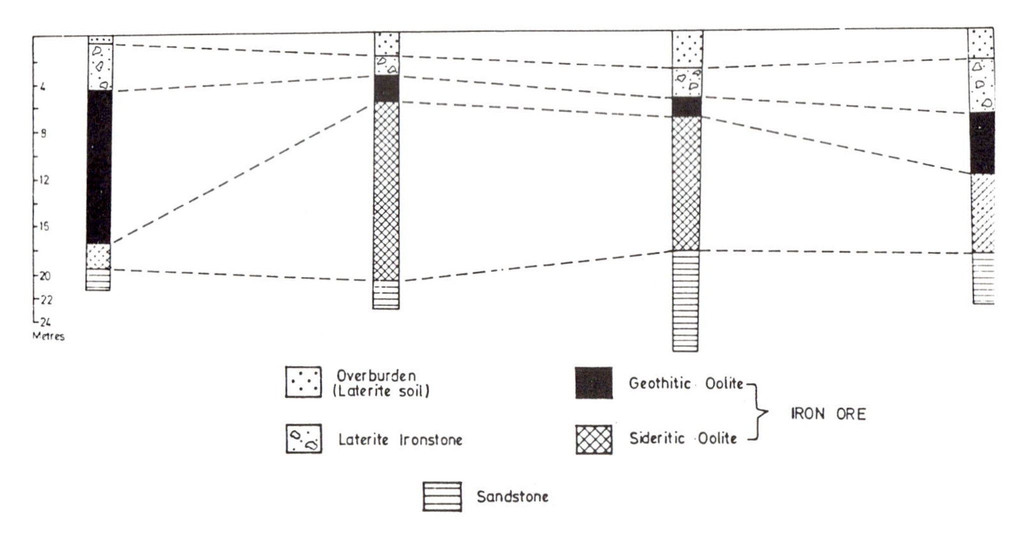

Figure 6. Stratigraphic section of oolite-pisolitic ironstones, Agbaja.

covering an area of approximately 4.6 sq km, reserves of ironstone with ore-grade of 45 - 50 % were roughly estimated at 132 million tons. There are indications that large potential reserves of ironstones are still available in the Agbaja Plateau and other parts of the Lokoja areas.

There are three textural varieties of the ore, which include the pisolite, fine-grained oolite and course-grained oolite (Fig. 7). It is the course-grained oolite that predominates. The Agbaja ironstone like its analogue elsewhere in the world is limonitic, geothite being the most common ore occurring as ooliths, pisoliths and matrix. Siderite is abundant in the weathered ironstone but it is largely restricted to the base of the oolite section. The bulk of the ironstone consists of brown goethite oolite, subordinate grey or black sideritic oolite and minor hematite and maghemite.

Variable amounts of phosphorus and aluminium are present in the Agbaja ironstone. Chemical analysis of the ironstone indicates that its phosphorus content varies from 1.73 - 3.00 % and alumina from 8.16 - 14.74 %. It is on account of the relatively high values of these elements which are deleterious to Fe metallurgy, that the ironstone deposits have not been exploited.

In the Sokoto basin the limonitic oolite ironstone occurs as part of the Paleocene Sokoto Group of sediments (Kogbe, 1978). The dominant lithologies in the sequence consist from bottom to top of basal bluish grey shale (Dange Formation), shaly limestone (Kalambaina Formation), upper shale (Gamba Formation), oolitic ironstone and laterites (Fig.8). The uppermost laterites varying in thickness from 3 - 5 m were derived from the oolitic ironstone by lateritic weathering which also affected other sediments of the basin. The unweathered basal oolite section has a thickness of 4 - 5 m. Goethite is the main ore mineral of the ironstone but subordinate hematite is present too. Quartz, silty clay and calcareous materials occur either in form of nuclei or matrix in the oolite. According to Kogbe (1978) the oolitic ironstone originated in a regressive marine to brackish water environment. This is however contrary to the origin proposed by Adeleye and Dessauvagie (1972) and Adeleye (1973) for the oolitic ironstones of the Niger basin dicussed above.

So far the full potential of the Sokoto ironstones is yet to be evaluated. However, Kogbe (1978) has reported that the total Fe content of the oolites is in the range of 60 -70 %. Chemical anaylsis of some specimens of the oolitic ironstone has indicated a noteworthy concentration of Mn, V and Ti.

d) Laterites

The surficial iron-enriched weathering products described as laterites are of widespread occurrence in Nigeria. They are derived from both crystalline and sedimentary rocks alike. The laterites can be classified into three types which occur at different physiographic levels.

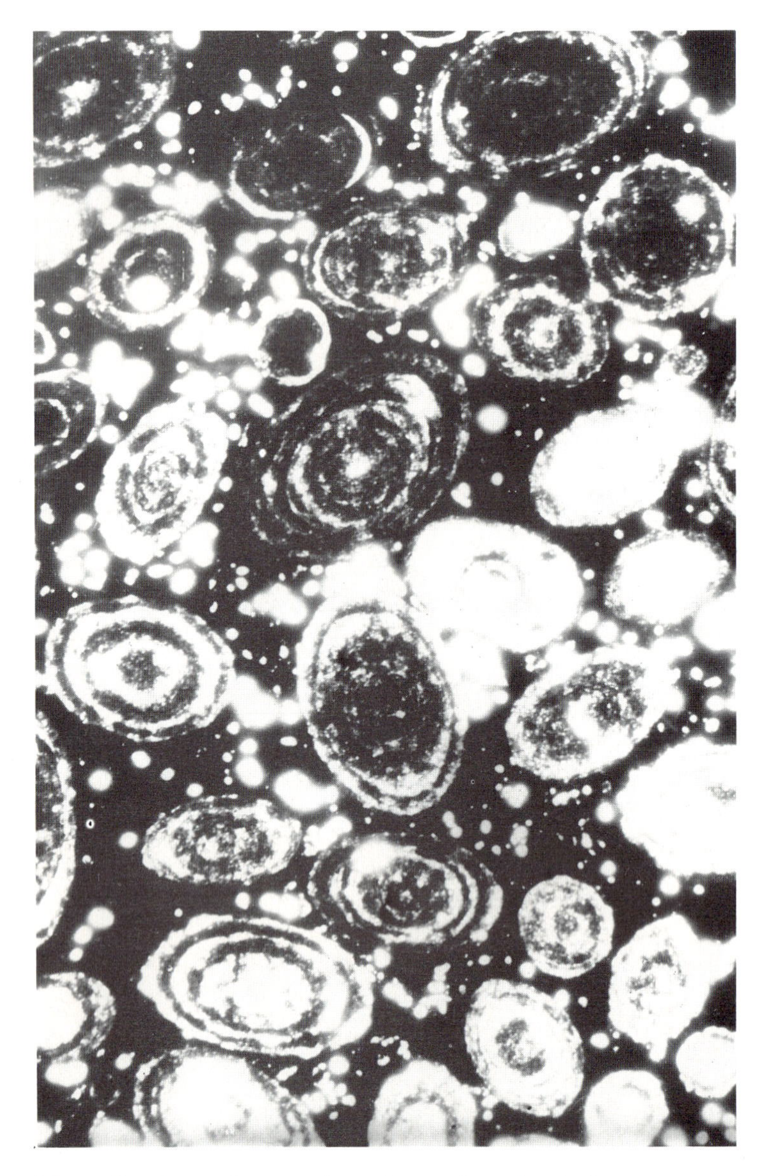

Figure 7. Photomicrograph of thin section of Agbaja iron ore.

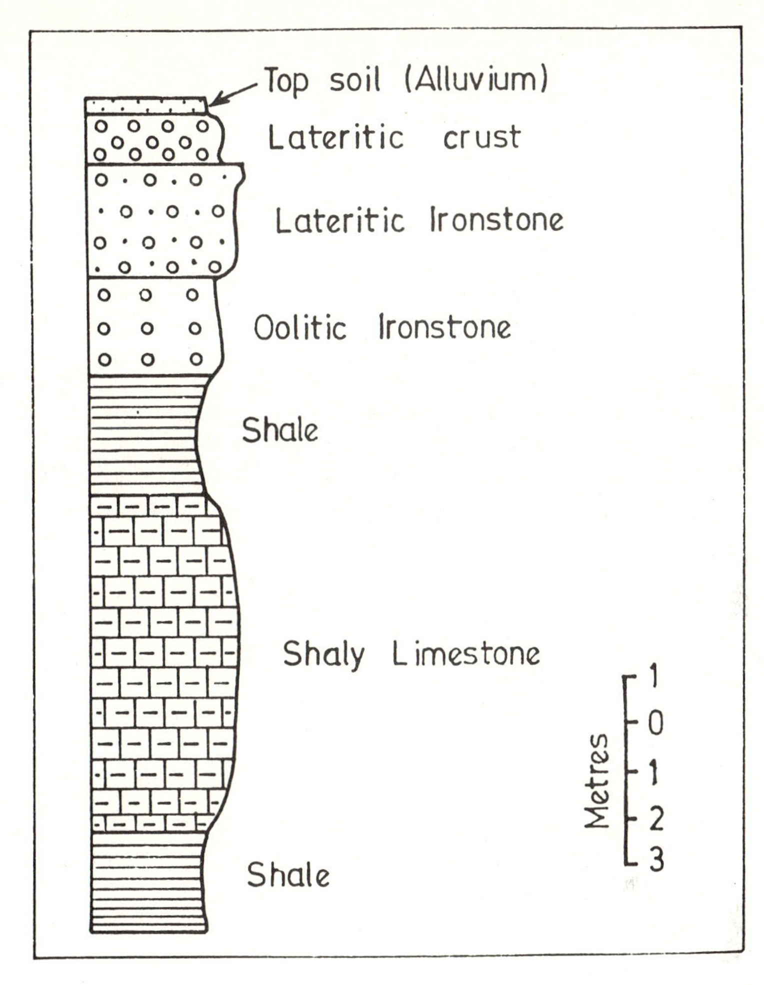

Figure 8. Stratigraphic section of the Sokoto Group showing the oolitic and lateritic ironstones (After Kogbe, 1978).

The first type is known as the peneplain or high-level laterite (du Preez, 1956) which was formed in Mid-Tertiary on an extensive peneplain but is today reduced to remnants. The laterite is now found as flat-topped cappings on picturesque mesas and hills, notably, at Jos Plateau and Benue, Kano, NIger, Kwara, Kaduna and Oyo States. Occurring at lower erosional levels are the second and third types of laterites. The second is usually found in the weathering profile and is usually exposed where the overlying top soil is washed off as at the breaks-in-slope. Such laterite is variously described as lower.lying illuvial laterite (du Preez, 1956); etch-plain laterite (Thomas, 1969); and pediment laterite (Folster, 1969). The third variety is the detrital laterite derived from the pre-exisiting laterites by erosion and deposition.

Of the three laterite types only the peneplain and detrital laterites offer some potential as an iron resource. Unfortunately the study of the Nigerian latrites in this regard is very much limited. The lateritic ironstone which occurs on the Udi Plateau west of Enugu and the laterite oolite overlaying the sedimentary sideritic ironstone of the Agbaja plateau (Fig. 1) are the only known lateritic iron deposits investigated in some detail (Hazel 1958; Jones 1958). While the Enugu ironstone is a detrital laterite the Agbaja oolitic laterite is an in-situ peneplain laterite.

The Enugu ironstone was derived from the lateritized surface beds (shale and sandstone) of the Upper Cretaceous Nsukka Formation which formerly covered the Udi Plateau. The laterite was eroded and deposited on the lower slopes formed by the underlying Ajali Sandstone Formation of the Enugu escarpment. The ironstone occurs in a bed of 0.6 m to 8.4 m thick, which is overlain by red clayey sand of 9.3 m average thickness. It is composed of unconsolidated angular laterite fragments of varying dimensions in a clayey sand matrix. Low grade iron ores of about 60 million tons with a tenor of 31.9 %, were proven near Enugu. Further beneficiation of the ironstone to 40 % Fe reduces the ore tonnage to 40 million. The rubbly Enugu ironstone is limonitic and the main ore mineral is goethite. However, minor amounts of hematite, pyrite and siderite have also been found. Quartz and clay are the main gangue minerals present. Texturally, the lateritic ironstone is mainly concretionary and cellular.

During a bauxite prospecting exercise in 1978 on the Worku Plateau, near Oju, SE Benue State, a shale-derived peneplain laterite capping (Fig. 1) was discovered to have 1.8 - 9.0 m thickness with an average of 5.9 m and 11 - 79 % Fe with a mean value of 30 % Fe (Adekoya et al., 1978). The plateau area covered by the laterite is about 2 sq. km. The laterite should have been considered to be a low grade iron ore but for its relatively high values of Al_2O_3 and SiO_2 which are metallurgically deleterious. However, the deposit still needs to be fully investigated before its full potential can be properly evaluated.

Laterites derived from iron-rich amphibolites have been mined locally in several areas particularly around Ilesha in S.W. Nigeria.

CONCLUSION

From the foregoing review it is evident that Nigeria is endowed with a considerable iron ore potential which, if properly explored and exploited, can provide sufficient raw material base for the manufacture of iron and steel which are required for the nations industrial development.

The ferruginous quartzites of the Okene-Lokoja district in Kwara State are at present the most important national source of iron ore. The largest of these deposits are the Itakpe iron ore bodies which are currently being developed to support the Ajaokuta Steel Plants now at an advanced stage of construction. With the estimated reserves of over 300 million tons and assumed mean annual production of 4 million tons the Itakpe deposit could last at least 70 years. Given the right economic climate, the exploitation of both the Ajabonoko and Choko-Choko deposits could further extend the life span of the iron mining industry in the Okene district.

Preliminary available geological information suggests that the banded iron-formations located mostly in the northwestern and central parts of Nigeria also constitute a potential iron resource, the utilization of which has to await more detailed investigations and perhaps better economic times as they appear to be generally low-grade and small deposits.

Although the sedimentary and lateritic ironstones, form a substantial source of iron ore, their poor quality (in terms of deleterious components) does not encourage exploitation. It is hoped that the low-grade deposits can still be used either through beneficiation or blending with richer or higher grade ores. So far, not much attention has been focussed on the study of the ubiquitous Nigerian laterites as a possible source of iron ore. There is therefore an urgent need to conduct a nation-wide evaluation of these surficial deposits before most of them can be utilised as an iron resource.

REFERENCES

Adekoya, J.A., 1981: Precambrian iron-formation of Northwestern Nigeria. In Precambrian Geology of Nigeria, Vol. 1 in Press, Geological Survey of Nigeria, Kaduna.

Adekoya, J.A., 1981: B.C. Irokanulo and K. Ladipo, 1978: Report on the investigation for Bauxite at Worku Hills, near Oju Benue State. Geological Survey of Nigeria, Kaduna.

Adeleye, D.E., 1973: Origin of Ironstones, an Example from the Middle Niger Valley, Nigeia - Jour. Sed. Petrol. Vol. 43. No. 3, 709-727.

Adeleye, D.E., 1973: The Geology of the Middle Niger Basin. In Geology of Nigeria, Ed. C.A. Kogbe, 283-287. Elizabethan Publishing Co., Surulere, Nigeria.

Adeleye, D.E. and T.F.J. Dessauvagie, 1972: Stratigraphy of the Niger Embayment near Bida, Nigeria. In African Geology, Eds. T.F.J. Dessauvagie and A.J. Whiteman, 181-186, University of Ibadan, Ibadan, Nigeria.

Baer, H.P., 1982: Late Precambrian Submarine Volcanism in NW. Nigeria (Kibaran Cycle). Abstracts of the Proceedings, 41-42, 18th Annual Conference of NMGS, Kaduna, 1st - 5th March, 1982.

du Freez, J.W., 1956: Origin, Classification and Distribution of Nigerian Laterites. Proceedings 3rd International West African Conference, Ibadan, 1949, 223-234.

Folster, H., 1969: Slope Develpoment in Nigeria Late Pleistocene and Holocene. Göttingen Bodenkundliche Berichte, 10.

Hazell, J.R.T., 1958: The Enugu Ironstone, Udi Division, Onitsha province. Rec. Geol. Survey Nigeria, 1955.

Kogbe, C.A., 1978: Origin and Composition of the Ferruginous Oolites and Laterites of North-Western Nigeria. Geologische Rundschau Band 67, Heft 2, 662-674.

Olade, M.A., 1978: General Features of a Precambrian Iron Ore Deposit and its Environment at Itakpe Ridge, Okene, Nigeria. Trans Institute Min. Metall. London (Sect. B: Appl. Earth Sci.), 87, 81-89.

Park, C.F., 1974: Earthbound: Minerals, Energy and Man's Future. Freeman, Cooper and Company, 279.

Lateritic Iron Deposits in Anambra State Nigeria

L. P. Orajaka and B. C. E. Egboka

Department of Geological Sciences, Anambra State University of Technology, P.M.B. 0116, Enugu, Nigeria

Keywords: Anambra State, Iron Ore Deposits, Lateritization, Ferruginization

ABSTRACT

Lateritic iron ores occur in the following localities inAnambra State, Nigeria: Nsude, Ngwo-Uno and adjoining villages near Enugu, Oji River, Isiagu in Awaka, Owerre-Ezukala and in many other localities. They are concretionary and nodular and comprise mainly of redish to brownish iron oxides mainly geothite and limonite. The ironstones, which are concentrated in the upper part of the Ajali Sandstone and the lower part of the Nsukka Formation were derived from the overlying ferruginous Nsukka Formation. The iron form may have been transported in the form of ferrous oxide which was subsequently subjected to intense oxidation hydration and precipitated as $Fe_2O_3.3H_2O$. The tenor ranges from 35 to 80 percent iron oxide (Fe_2O_3). The iron ores contain low phosphorous but medium to high silica and alumina. The reserve is tentatively put at about 100 m tons.These deposits can no doubt support a viable iron and steel industry or a big rolling mill.

INTRODUCTION

Some previous workers such as Hazell (1955), and Orajaka (1972 and 1973) have written about the iron deposits in the area now called Anambra State.Their work showed that lateritic iron ore is found in Nsude, Ngwo-uno and adjoining villages near Enugu. Lateritic ironstones consisting of concretionary and nodular ironstones are seen on the road cutting on the Enugu-Onitsha Expressway in Udi and Oji local government areas and at cliffs in many parts of Nsukka, Udi, Enugu, Oji River, Awka and Aguata. The commercial importance of the Enugu iron ore is still undecided. The tenor has been given to range from 29 percent to 45 percent (Hazell, 1955). The Geological survey of Nigeria (1954) has indicated that nearly 50 million tons of iron are present around Enugu. The areal extent of the iron deposits, their quality and reserves are not yet established. Iron smelting and blacksmithing has been the main occupation of Awka people. The source of the iron used by the Awka blacksmiths has not been documnented. Preliminary investigations reveal that the source of the iron used by the Awka blacksmiths is from Isiagu area. Samples of iron ore from Isiagu were collected to determine their tenor. Geological studies at outcrops were undertaken to determine the approximate thickness of the iron ore.This investigation is carried out to determine the areal extent of the sedimentary iron ore, and to determine the quality and possible reserves of the iron ore. Samples of the ore collected from different localities in the state were analysed for their iron content.

PHYSIOGRAPHY AND CLIMATE

The study area consists of lowlands and hills, many of the hills especially those found toward Nsukka are rounded and flat topped. The hills are covered by grass (Fig. 1). From Nsukka to Nineth Mile Corner the hills get fewer toward the south. These hills are capped by a mixture of laterite and ironstone concretion and are outliers of the Nsukka Formation.The most prominent geomorphic feature in the area is a N-S trending cuesta whose east-facing escarpment rises to between 200 to 350 m above the Cross River Plain. Its dip slope is generally southwestward. It is about 440 km long, extending from the Benue valley in the north to Arochukwu in the south (Fig. 2). In plan, it has the shape of an inverted 'S' which is over 80 km across between Idah and Okaba narrowing to 40 km in the vicinity of Nuskka and then to 20 km in Udi. It tapers futher to 12 km at Okigwe before swinging east and then south to its terminus near Arochukwu where it is only 5 km across (Umeji 1980). On the dip slope of the cuesta are numerous residual hills and ridges. The residual hills are the remnants of the Nsukka Formation which form outliers of various shapes on the Ajali Sandstones. Between these hills and ridges are lowlands and steep walled, flat-bottomed valleys. In Awka, Njikoka and Aguata areas the landscape is dominated by a north-south cuesta called the Awka-Orlu upland. The eastern margin forms the scarp-slope while the weastern margin forms the dip-slope. Both the scarp-slope and dip-slope have been ravaged by erosion and gullying although the scarp-

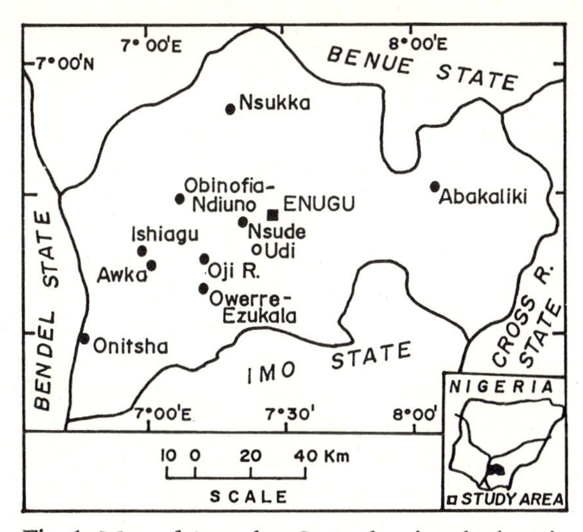

Fig. 1. Map of Anambra State showing the location of iron ore.

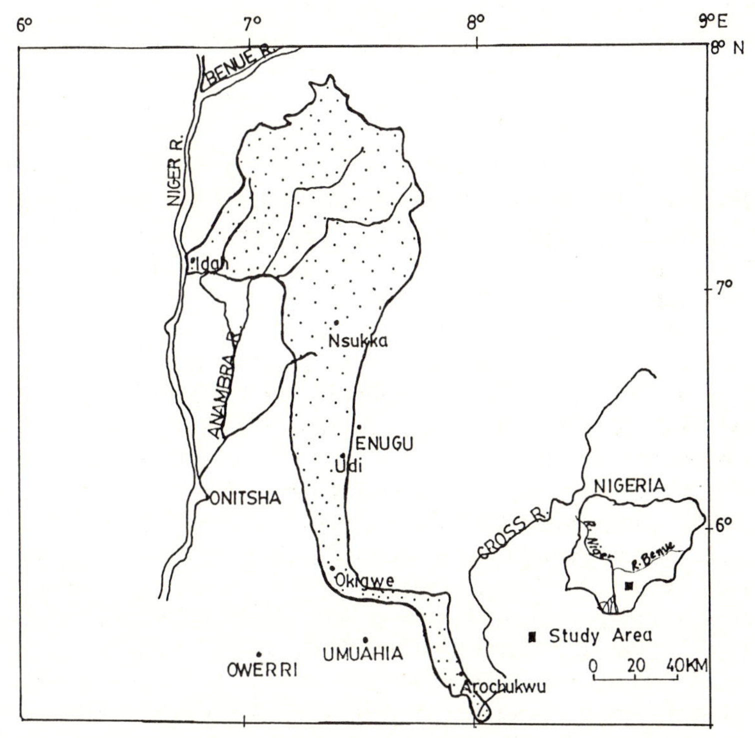

Fig. 2. Map of part of Southeastern Nigeria showing the cuesta and the major watershed.

slope has been gullied more intensly. The crest of the cuesta stands at over 378 m above sea level in many places.

The study area lies within the humid tropical rain forest belt of Nigeria. The annual rainfall in the area is over 2000 mm (Ogbukagu, 1976). Much of the rain falls in the months of May to October. The dry periods are characterised by very high temperatures and lower relative humidity while the wet months have lower temperatures and high relative humidity. Temperature is high most of the year except during the Hamattan (December - January). The month of December has low temperatures (20^{0} C) during the nights and mornings and warmer days (up to 30^{0} C). The average temperature for the year is about 30^{0} C. The hottest period is between February and early April.

Two main seasons exist in Nigeria, namely, the dry season that runs through the months of November to April and the rainy season that begins in May and ends in October. Floyd (1965) gave an average rainfall of 2032 mm for the areas now called Anambra State.

Groove (1951) showed that as much as 5088 mm of rain may fall in one month especially in June, July and September when strong storms are most prominant. The rainfall may occur as violent downpours accompanied by thunderstorms, heavy flooding, soil leaching, extensive sheet outwash, groundwater infiltration and percolation. The above climatic conditions, high rainfall in combination with high humidity and high temperature, common in the area is very significant in laterite formation. Other conditions for laterisation especially intense leaching, good drainage, strong oxidising environment and long durtion of weathering are also met in southern Nigeria.

METHODS OF STUDY

The distribution of the iron bearing formations was determined during fieldwork. Formation outcrops were examined at cliff, valley walls and road cuttings where iron ores are exposed along the newly constructed Enugu-Onitsha Expressway and the Enugu-Nsukka road. Samples of the ore collected from Nsude, Ngwo-uno, Isiagu. Oji River, Obinofia-Ndiuno, Owerre Ezukala and at many locations along the Enugu-Onitsha Expressway were analysed for their iron content using chemical method.

About 55 gm representative samples were collected at each sample location. Only clean and fresh samples with surface alteration removed were used. About 100 gram from each sample was pulverised and sieved through a 500 mesh screen to achieve a uniform size. 100 mm of the prepared sample was dissolved in 4 mls dilute HNO_3, boiled for about one hour and allowed to cool and filtered. The filterate was basified with dilute NH_4OH. Brown precipitate of iron oxide (Fe_2O_3) is formed. The

Table 1 Iron content of representative samples from parts of Anambra State.

Sample No	Location	% Fe_2O_3	Sample No	Location	% Fe_2O_3
101	Isiagu	69	111	Obinofia-Ndiuno	48
102	"	73	112	"	57
103	"	75	113	"	64
104	"	82	114	"	73
105	"	64	115	"	55
106	Nsude	49	116	Owerre Ezukala	38
107	"	53	117	"	33
108	"	45	118	"	36
109	"	38	119	"	27
110	"	35	120	Oji River	29
			121	"	34
			122	"	37
			123	"	35

Table 2 Geologic age and lithostratigraphic sequence in Anambra State.

GEOLOGIC AGE (MILLION) (YEARS)	LITHOLOGIC SEQUENCE
MOICENE (26)	Benin Formation - - - - - - - - - - Lignite Series
EOCENE (53.4)	Ameki Formation Nanka Sands Formation
PALAEOCENE (65)	Imo Shale Formation (Ebenebe Sandstone)
MAESTRI-CHTIAN (70)	Nsukka Formation Ajali Sandstone Mamu Formation
CAMPANIAN (76)	Enugu Shales Formation Awgu Shales Formation (Awgu Sandstones)

precipitate was filtered out, dried and weighed. Care was taken to avoid any loss. Percent iron oxide (Fe_2O_3) was calculated using the formular:

$$\% \text{ Fe (as } Fe_2O_3) = \frac{\text{weight of precipitate}}{\text{weight of sample}} \times \frac{100}{1}$$

The results of the analyses of some representative samples are given as percent iron oxide (Fe_2O_3 %) in Table 1.

Geologic Setting

The iron bearing strata lie within the Maestrichian Ajali Sandstone and Nsukka Formation.The stratigraphic succession in the study area is listed in Table 2. The geologic map of parts of Anambre State is given in Figure 3. The Nsukka Formation which succeeds the Ajali Sandstone is well exposed in the valley of the Nadu River a few kilometres north of Nsukka. It occupies a broad stretch of land west of Udi Plateau from Nsukka towards Onitsha (Figure 3). Greater parts of the Udi Plateau surface is formed over the Ajali Sandstone but the masses are found on the Nsukka Formation (Umeji 1980). The steep valley sections which are more than 100 m deep, often expose the contact between the overlying Nsukka Formation and the Ajali Sandstone. South of Enugu-Ontisha road the outcrop is about 12 km wide but futher north it broadens as a result of the decrease in regional dip. The Nsukka Formation consists of coarse sandstones with shale intercalation and fragments of nodular iron stones and ferruginised indurated shale and sandstones. Differential erosion has left the resistant portions of the Nsukka Formation stand out as rounded, conical and sometimes flat-topped hills some hundreds of meters above the general level.

The Ajali Sandstone is exposed in Ajali River in Udi local government area. The sandstone is thick, friable and poorly sorted and is white in colour. There are iron stains which signifies ferruginisation and infact on the Udi Plateau, ferrugnisation of the formation is common. Pisolitic iron ore exposed along the Enugu-Onitsha Expressway is found in Ajali Formation. Bands of iron concretions and nodules are common in the Ajali Sandstones.

Distribution of Iron Ore and Reserves

The nodular and concretionary sedimentary ores occur in the following localities in Anambra State: Nsude, Ngwo-uno and adjoining villages near Enugu, in parts of Oji River, Obinofia-Ndiuno in Ezeagu, Owerre-Ezeukala in Aguata area, Isiagu in Awka and in many other localities (Fig. 1). The lateritic ironstones consisting of concretions and nodular bands within Nsukka Formation and Ajali Sandstone can be seen on the road cutting in the Enugu-Onitsha Expressway and at cliffs around Obinofia-Ndiuno, Isiagu and Owerre-Ezukala. The iron ore though extensive occurs

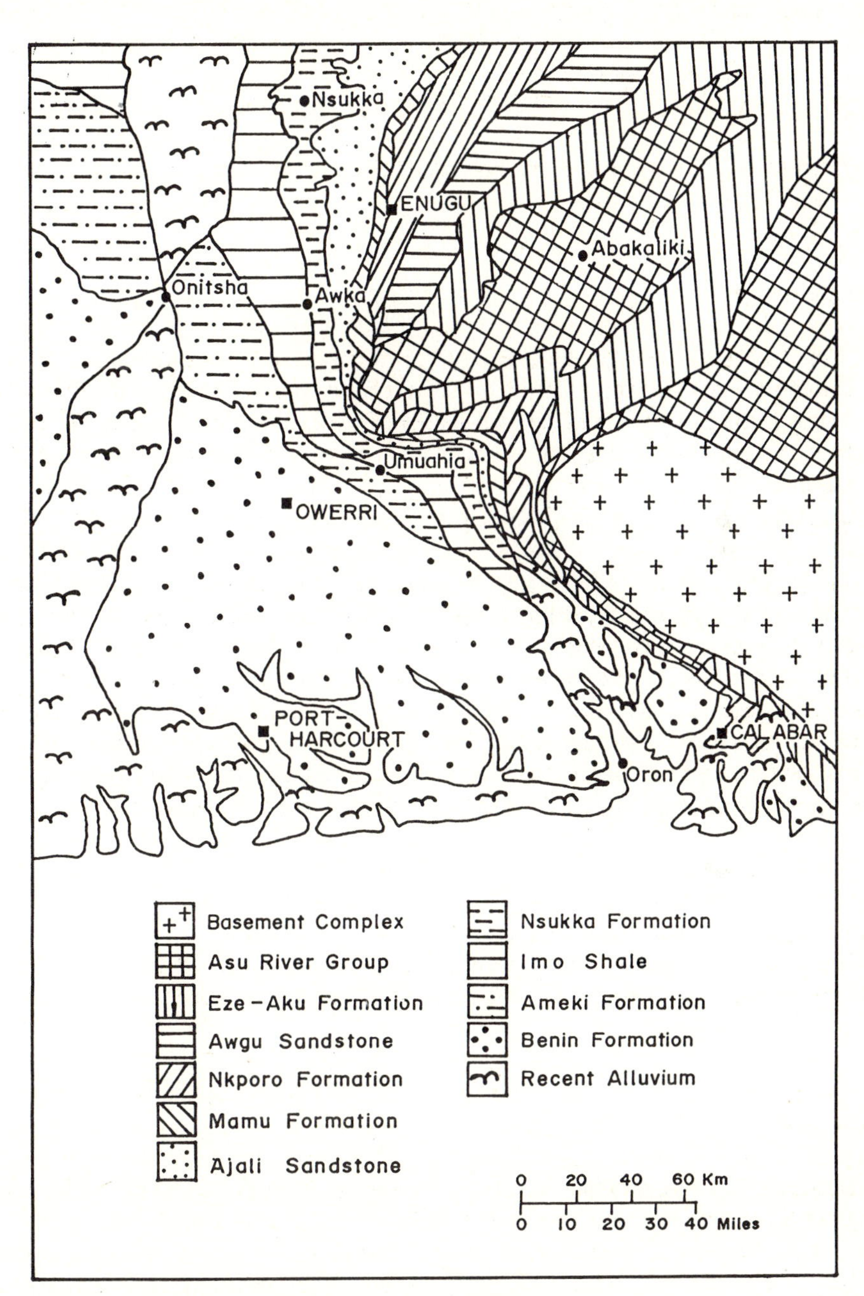

Fig. 3. Geologic map of Eastern States of Nigeria.

in thin bands averaging about 1.5 meters in many places in the upper part of the Ajali Sandstone and the lower part of the Nsukka Formations. Lateritic cuest and highly ferruginised soil reach depths of 10 - 15 m in many localities. The Geological Survey of Nigeria (1954, 1957) has indicated nearly 50 million tons of ironstone around Enugu. This estimate does not include the deposits discovered during this work at Obinofia-Ndiuno, Isiagu and Oji River. With these recent discoveries the reserves may double the estimate of the Geological Survey. Although the iron-stones occur in thin bands their large area extent makes them quantitatively important and economically exploitable.

Quality of the Iron Ore

The ironstones of Anambra State are concretionary and nodular and comprise of hematite, geothite and limonite. Hazell (1955) and Orajaka (1972) reported the tenor of the ironstones to range from 29 percent to 45 percent. The average iron content of 50 ironstone samples analysed in this work is 43 percent. Iron concretions that contain 75 percent iron oxide (Fe_2O_3) are present at Isiagu and Obinofia-Ndiuno. The deposit at Isiagu supported local blacksmithing industry. The reserves may be up to 100 tons. Some iron bands at Isiagu contain up to 80 percent iron (expressed as Fe_2O_3). The Anambra State iron ores contain low phosphorous, mediun silica and alumina (Orajaka, 1973). In areas where the ore are thick and continous for a good distance especially in Ngwo, Nsude, Isiagu the iron ore can easily be mined by strip mining methods in other areas the ironstone is not thick and therefore cannot be mined at a profit. It is recommended that a more detailed mapping and preliminary drilling be undertaken by prespective miners to delineate areas that contain economically exploitable deposit.

Genesis of Iron Ore

The ironstones of Anambra State occur in the lower part shale/siltstone units in the Nsukka Formation and in the Ajali Formation. They are concretionary and nodular and comprise mainly of geothite and limonite. They occur in thin bands which generally extend to few meters. These ironstones are believed to be derived from the overlying ferruginous Nsukka Formation (Hazell, 1955, Orajaka 1972). The ironstones are evidently lateritic in origin. The iron may have been leached and transported in the form of ferrous hydroxide which was subsequently subjected to intense oxidation, hydration and precipitation as Fe^{3+} hydroxide. The precipitated iron oxide occurs in bands in the zone of accumulation as hydrated ferric oxide.

SUMMARY AND CONCLUSIONS

The commercial importance of the iron deposits in Anambra State is as yet not fully determined. Because of limited resources a detailed drilling work necessary for the determination of accurate reserve figure was not done.Futher detailed work is necessary for a more reliable reserve determination. The Geological Survey of Nigeria (1954) indicated that nearly 50 million tons of iron ore present around Enugu when the new discoveries in Obinofia-Ndiuno, Isiagu and Oji are added to the Enugu iron ores the reserves of ironstone in Anambra State may be as much as 100 million tons. The Anambra State ores contain low phosphorous but medium silica and alumina. They are also close to the Enugu coal deposits and are not far removed from limestone deposits in Nkalagu. The Enugu coal is non-cooking but the coal mixed with a portion of good quality cooking coal in Britain has proved to be good for iron smilting (Orajaka, 1979). These deposits can support a viable iron and steel industry or a big rolling mill. The geologic, environmental and economic conditions are highly favourable for the establishment of iron ans steel industry.

ACKNOWLEDGEMENTS

The assistance of Mr. M.M. Irogbunachi at various stages of this investigation and preparation of this paper is acknowledged. Sincere thanks, are due to Miss Chinyer Nwolisa for typing the script and to Mr. F. Ozoani for drawing the illustrations. The financial support of Anambra State University of Technology, Enugu is acknowledged.

REFERENCES

Egboka, B.C.E. and G.I. Nwankwor, 1985: The hydrogeological and geotechnical parameters as agents for gullytype erosion in the rain-forest belt of Nigeria. Jour. African Earth Sci., Vol. 3, No. 4, 417-425.

Floyd, B., 1965: Soil erosion and deterioration in Eastern Nigeria. Nigerian Geor. Jour. Vol. 8, 33-44.

Groove, A.T., 1951: Land use and soil conservation in parts of Onitsha and Owerri provinces Nigeria. Geol. Survey of Nigeria, Bull No. 21.

Hazell, J.R.T., 1955: The Enugu Ironstone, Udi Division, Onitsha Province. Record, Goel. Surv. Nigeria, 44-58.

Ogubukagu, IK. N. 1976: Soil erosion in the Northern parts of Awka-Orlu upland, Nigeria. Jour. Mining and Geology, Vol. 13, 6-19.

Orajaka, I.P., 1979: Economic Geology of Coal deposits in Nigeria. Unpubl. Company Report, Mapco Inc., Tulsa, Oklahoma, U.S.A. 18.

Orajaka, S.O. 1973: Possible metallogenic provinces in Nigeria. Econ. Geol., Vol. 68, No. 2, 278-280

Orajaka, S.O., 1972: Nigerian Ore deposits. Mining Magazine, October, 357-359.

Umeji, A.C., 1980: Tertiary plantation surfaces on the cuests in southeastern Nigeria. Nig. Jour. Mining and Geol., Vol. 17, No. 2, 109-117.

Annual Report of the Geological Survey Department of Nigeria 1953 - 1954. Federal Government Printer, 13-15.

Mineral and Industry in Nigeria, 1957. Federal Government Printer, 23-25.

On Determining Weathered Layer Velocities and Depths to the Lignite Seams of the Anambra Basin, Nigeria by Uphole Seismic Reflection Method

P. O. Okeke and L. N. Ezem
School of Natural & Applied Sciences, Federal University of Technology, P.M.B. 1526, Owerri, Nigeria

Keywords: Coal and Lignite exploration, Uphole seismic reflection, Anambra Basin, Seismic velocities.

ABSTRACT

Data of weathering parameters (weathering depth (Dw), weathering velocity (Vw), and sub-weathering layer velocity (Vc)) from an uphole seismic reflection shooting experiment in Onitsha region of lignite series of the Anambra Basin, Nigeria are presented. Dw ranges from 16 to 24 metres with a regional representative value of 20 metres, Vw from 430 to 600 m/sec. (regional representative value of 500 m/sec.) and Vc from 1000 to 2500 m/sec. (regional representative value of 2000 m/sec.). Results suggest that Dw can be determnied using information from all the 24 geophones and not necessarily from only the conventional first (nearest) uphole geophone. Becsuse of interplay of reflection and refraction effects, Vw can best be determined from the first three geophones and the last five, while the intermediate geophones (4 to 19) give erroneous results. Also, since velocity of reflected wave tends to vary with change in depth, the slope of the resulting parabolic curve may not be accurately determined as to be used in calculating Vw. Over-all, best results seem to be obtained from the first three geophones.

INTRODUCTION

As part of seismic exploration survey, uphole reflection shooting is usually carried out in any region in order to determine average weathering depth, average weathering velocity and average sub-weathering layer velocity. The results so obtained are then used for detailed in-line seismic reflection shooting by split-spread or any other selected geophone - shot-point arrangement.

An uphole survey is one of the best methods of investigating the near-surface, and finding the thickness (Dw), velocity (Vw) and sub-weathering layer velocity (Vc). The survey basically involves two layers of the earth - the weathering layer and the sub-weathering layer. Usually a complete spread of geophones, including uphole geophone, is used. Shots are fired at various depths up the hole, with the uphole geophone planted about 5 metres 'offset' on the ground surface from the shot hole. A shot hole deeper than the base of the weathering layer is required and is usually cased to avoid caving. This makes for better results. One-way arrival times are recorded as the shots are taken at different depths.

Normally assumptions made in carrying out any detailed uphole survey weathering corrections include:

a) Vertical ray paths
b) Small 'off set'
c) Fairly constant Dw between uphole geophone and shot hole.
d) Availablity of uphole breaks, and
e) Shots taken below the weathered layer.

This study was aimed at determining seismic weathering layer velocities and depths and sub-weathering layer velocity in the Lignite seams of Lignite Formation (Upper Eocene) using information from 24 geophones instead of only the conventional first (uphole) geophone (Okeke, 1982).

The need for geophysics in the coal industry is mostly quite different from its need in the oil industry. Geophysics is used in the oil industry to help find oil reserves. Seismic reflection surveys in particular are used to locate promising places where oil and gas reserves may exist. Normally, many large areas are excluded from futher consideration because they hold no promise of either oil or gas. Coal (pure or impure/lignite) is not difficult to find by seismic methods (Zoilkowski, 1982).

The problem is to mine it cheaply, and geological structures may affect the costs of extraction. However, if indirect gravity evidence suggests the extent of the desired structure, and a seismic follow-up has sufficient resolution to pick up the structure, then seismic reflection should be applied to determine it. The British National Coal Board has used seismic Low Velocity Layer (LVL) refraction shooting to determine precisely the parameters discussed in this paper in North Wales (SS(E)L; 1978).

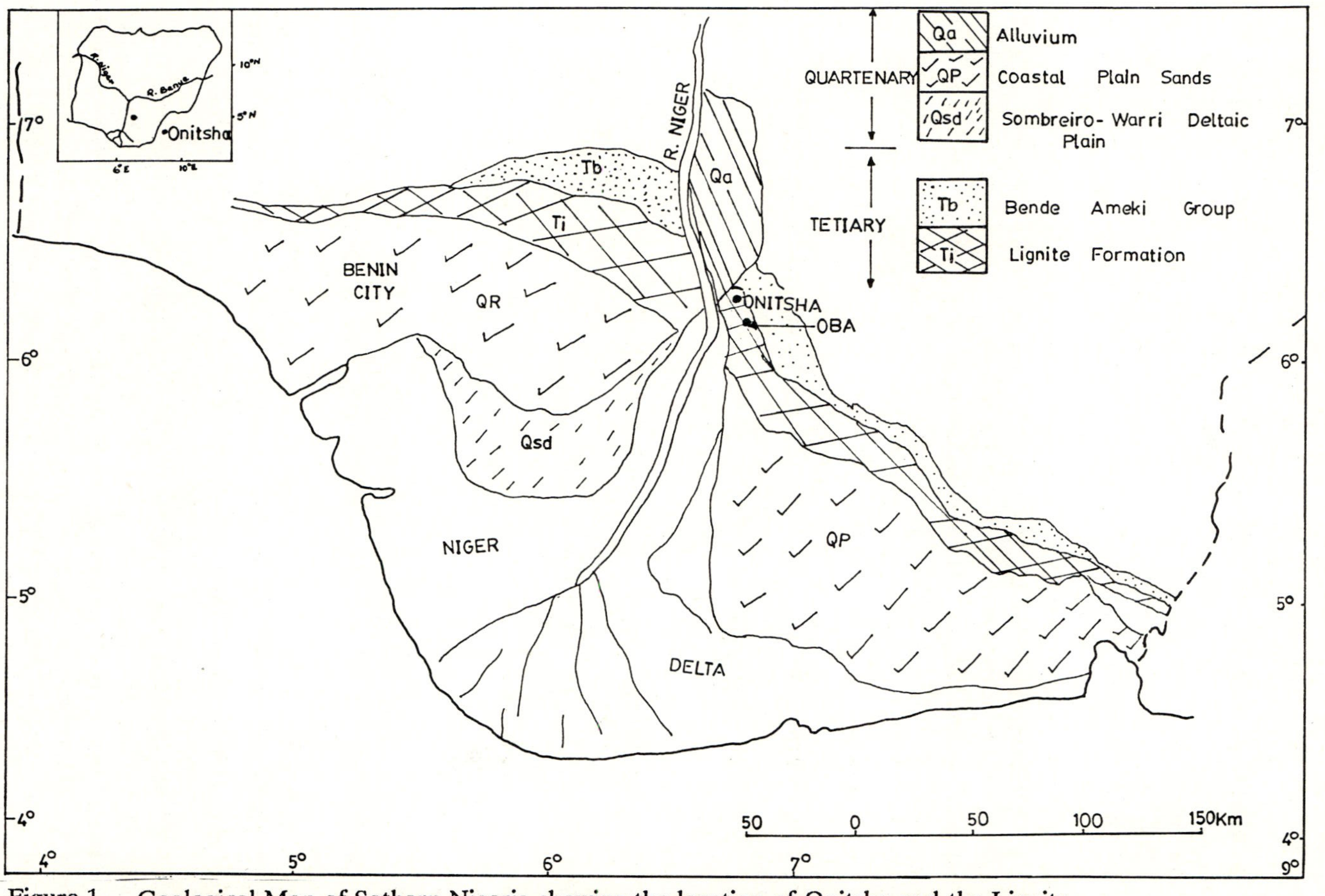

Figure 1. Geological Map of Sothern Nigeria showing the location of Onitsha and the Lignite Formation (Produced from Geological Map of Nigeria, 1974 edition).

Seismiograph Service (Nig.) Ltd. and the Nigerian National Petroleum Corporation (NNPC) have used the Plus and Minus method for the same purpose in the study region for oil exploration.

The selected area for this investigation is Onitsha within the Lignite series of the Anambra Basin. The Lignite series consist of alternation of lignite seams and clays that directly underlie the weathered and loose soil cover. Hence, in mineral exploration for shallow deposits such as the impure coal/lignite in Onitsha region, a knowledge of Dw should help in determining the drill depths.

LIGNITE DEPOSITS OF NIGERIA

All the known seams of lignite within the Anambra Basin of Nigeria occur west and east of the Niger (Fig. 1), the thickest seams being near Ogwashi-Asaba area. These areas have been explored by mapping and drilling. In the Ogwashi-Asaba area, there are two thick lignites, a main seam, averaging 6 metres in thickness, separated from an upper seam with a mean thickness of 3 metres by about 4 metres of clay-shale. The average ratio of overburden to lignite is 6:1, which appears to rule out open-casting (Federal Govt. of Nigeria Paper, 1957). Other lignite zones within the west of the Niger include Obomkpa, Ibusa, Okpanam, Illah, Agbor and Ubiaja. The most important seams east of the Niger are found near Oba and Nnewi. At Oba, near Onitsha, there is a lower and upper seam respectively 2 metres and 1 metre thick. This study was undertaken in the Oba-Onitsha axis. At Nnewi there are also two seams, the lower one 4 metres and the upper 1 metre thick. Lateral extension of these lignites is still being traced, but lignites have been reported near Orlu and Umahia both in Imo State of Nigeria.

EXPERIMENT

This study was organised in conjunction with the Nigerian National Petroleum Corporation (N.N.P.C.) Seismic Party 'X' based in Onitsha. The best results from the 24-channel recording came from shot point 690 (a cased hole) on seismic line 99 - 78 - 115 extension, and these have been discussed in this paper. The shooting details are given in Table 1. A total of 16 shots were taken up-the-hole. One-way time arrivals obtained at various depths have been used in the plots (Figs. 2.A - 2. J).

Shots were retaken at certain depths due to misharp and unclear first breaks, and necessary corrections were made for anomalous data. Reflection paths of the records were studied and the best results came from shots taken below the weathered layer. However, interferring waves like hole noise and groundroll were checked by burying all geophones properly and special shallow shots taken from a smeared cap. Casing - and low amplitude - breaks were also avoided.

Table 1: Shooting Details

Shot No.	DEPTH (metres	Amount of Charge Used
16	3	1 cap
15	6	$\frac{1}{3}$
14	9	$\frac{1}{3}$
13	9	$\frac{1}{3}$
12	12	$\frac{1}{3}$
11	12	$\frac{1}{3}$
10	15	$\frac{1}{3}$
9	15	$\frac{1}{3}$
8	15	$\frac{1}{3}$
7	15	$\frac{1}{3}$
6	18	$\frac{1}{3}$
5	20	$\frac{1}{3}$
4	25	$\frac{1}{3}$
3	28	$\frac{1}{3}$
2	33	$\frac{1}{3}$
1	38	$\frac{1}{3}$

NB: Uphole geophone offset = 5 metres

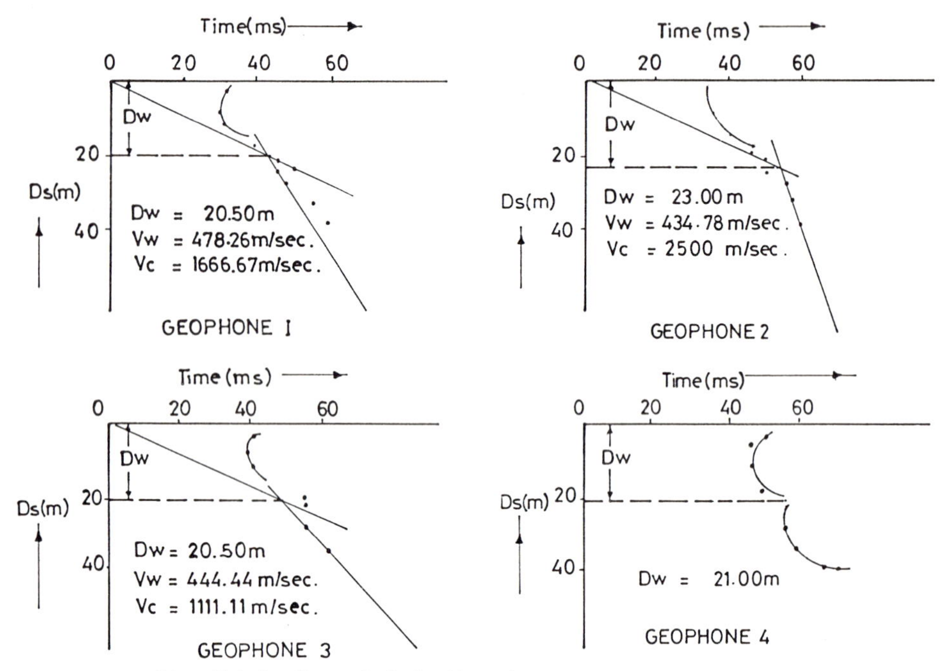

Fig. 2A. Uphole time-shot depth (Ds) plots for geophones 1-4.

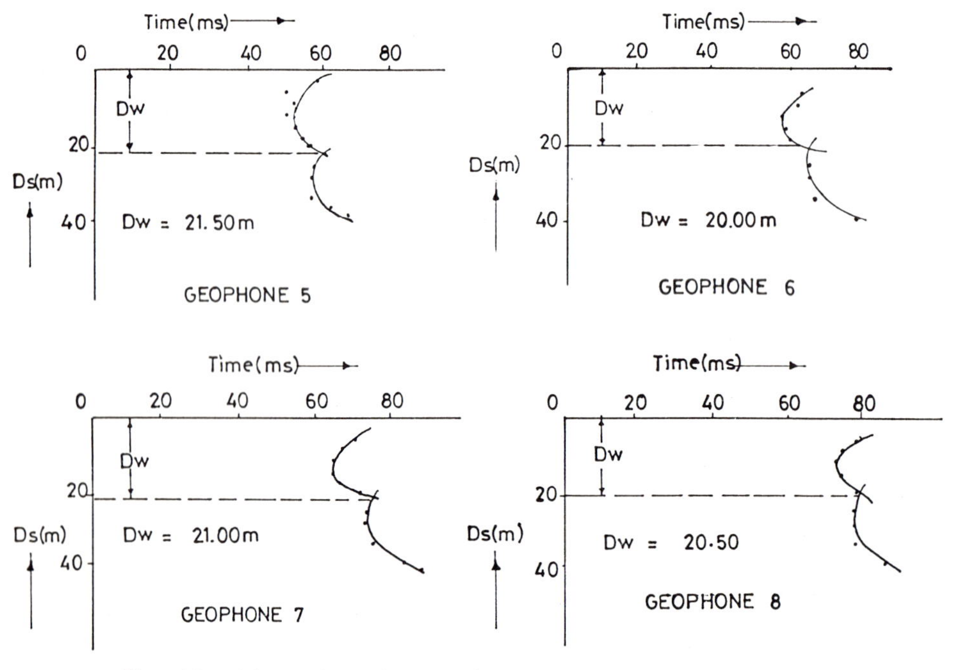

Fig. 2B Uphole time-shot depth (Ds) plots for geophones 5 - 8

Figures 2.A - 2.J Uphole time - Shot depth plots for the 24 geophones.

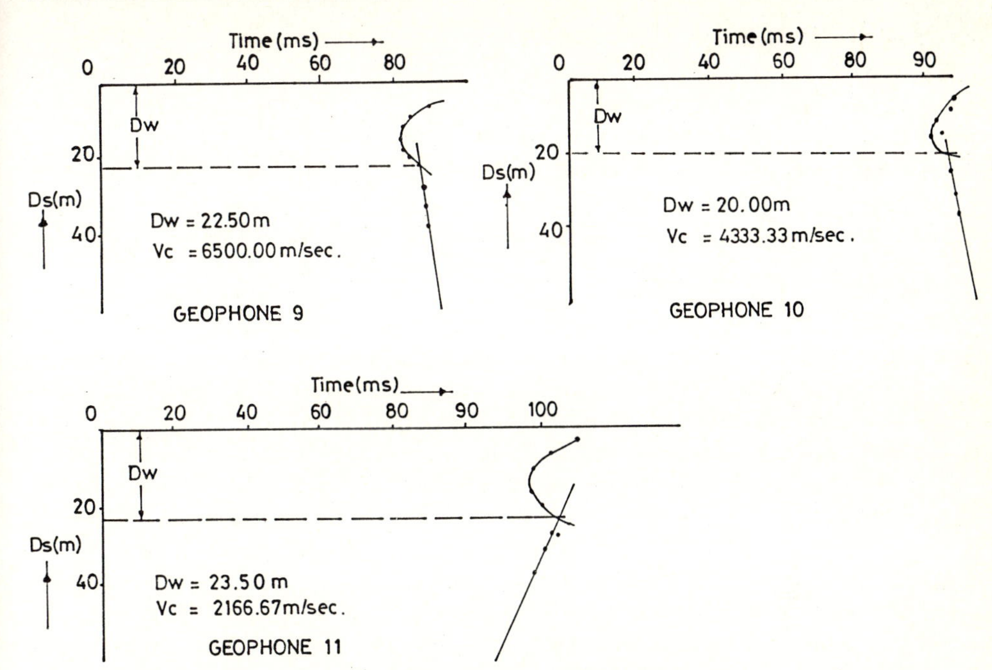

Fig. 2C. Uphole time-shot depth plots for geophones 9 -11

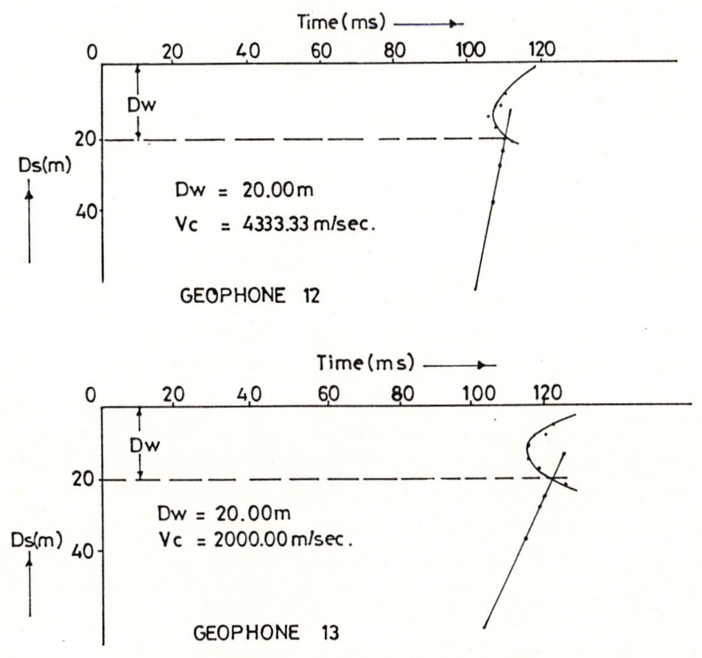

Fig. 2D. Uphole time-shot depth plots for geophones 12 and 13.

DATA INTERPRETATION

For a proper interpretation of uphole data, care must be taken in picking first breaks. A standard procedure on all surveys is to pick reflection times on two good reflections recorded from all shot depths on an end-on spread. The time break is then picked to the nearest half-millisecond. Changes in arrival time differentials between the nearer and more distant geophones show velocity changes or hole drift.

Arrival times have been plotted against shot depths for the first/uphole geophone and for several of the distant geophones. The plot for the first geophone changes abruptly where the shot enters the low velocity layer; the slope of the portion above the base of the weathering layer gives the weathering velocity (Vw) and the break in slope defines the depth of weathering (Dw) (Telford et al., 1976). The sub-weathering velocity (Vc) can also be obtained from the slope of the time - distance curve for the first arrivals on the reflection record. Figures 2. A - 2. J show these plots and the results so obtained. For the distant geophones the plot is almost vertical at first since the path length changes very little as long as the shot is in the high speed-layer. However, when the shot enters the weathering layer there is an abrupt change in slope and the travel time increases rapidly as the path length in the weathering layer increases. The reflection velocity which approximates to 1.1 Vc at the base of the weathering layer, is obtained by dividing the time interval between the vertical portions of the curves for two widely separated geophones into the distance between the geophones if known. The velocity measuremnt is often different from that given by the slope of the deeper portion of the uphole geophone portion curve, partly because the layering of beds of different velocities has little effect on the time between two widely separated geophones, but may affect the uphole time substantially (Telford et al., 1976).

Reflection time is plotted against depth of shot to determine the compatibility of reflection time differences to uphole time differences between shots. Deviations from a straight line on this plot may be due to multiples, improper picking of first breaks or some other anomalous conditions.

The graphs (Figs. 2. A - 2. J) show that the geophones nearest to the shot hole (including the first/uphole geophone) receive direct one-way arrivals and since, in principle, an uphole geophone is place on top of or very close to the shot hole, the graph in the first layer passes through the origin, and is a straight line. Another straight line graph is obtained in the second layer. These two lines intersect at the interface that defines the weathering depth (Dw). The slope of the line for the first layer gives Vw, and Vc is slope of the line in the sub-weathering layer.

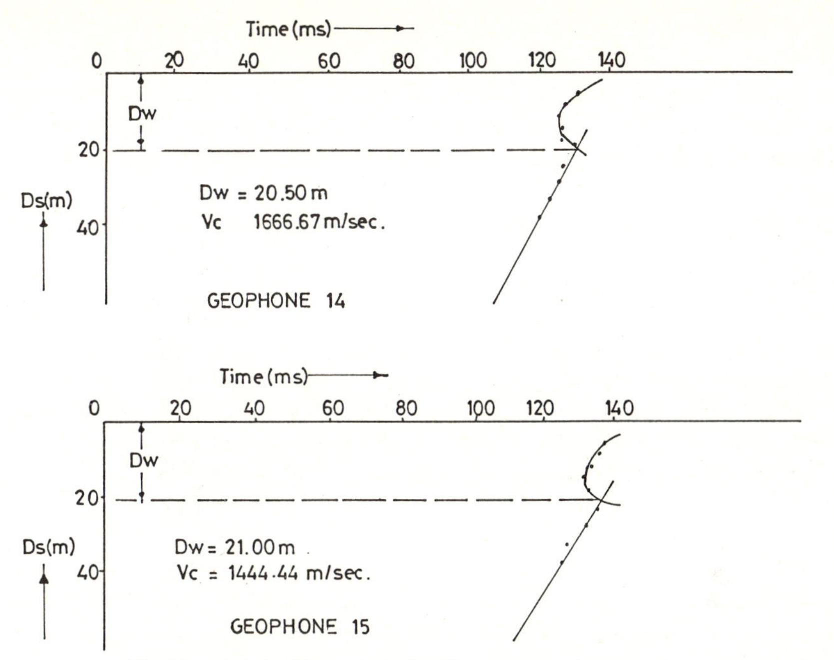

Fig. 2E. Uphole time-shot depth plots for geophones 14 and 15.

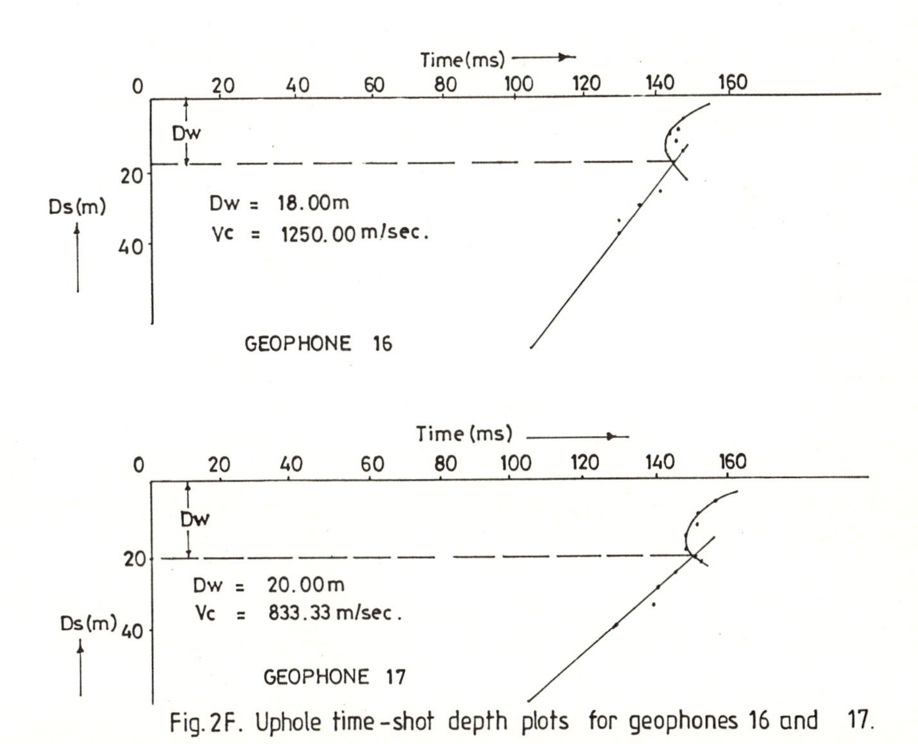

Fig. 2F. Uphole time-shot depth plots for geophones 16 and 17.

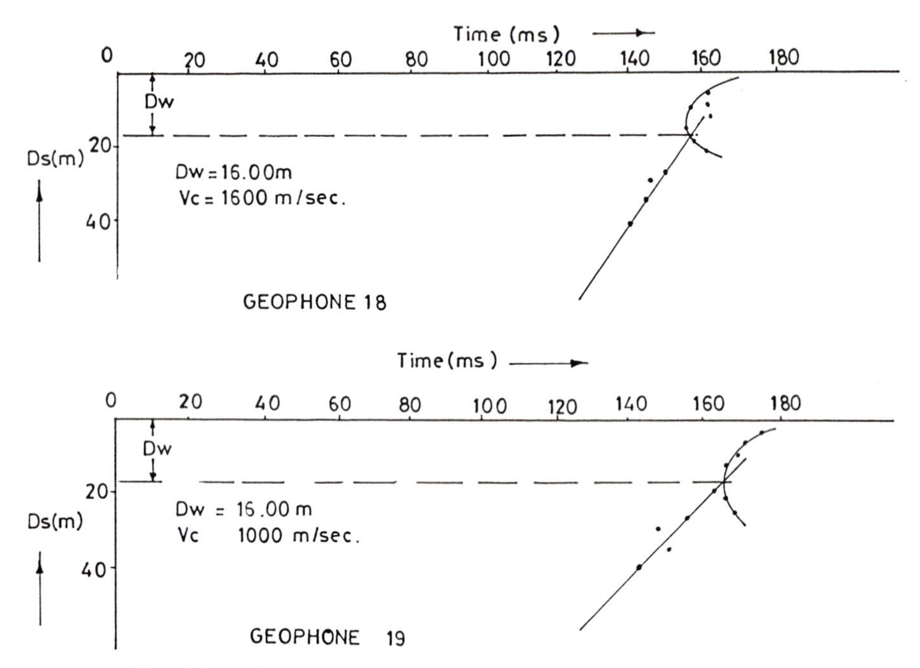

Fig. 2G. Uphole time-shot depth plots for geophone 18 and 19

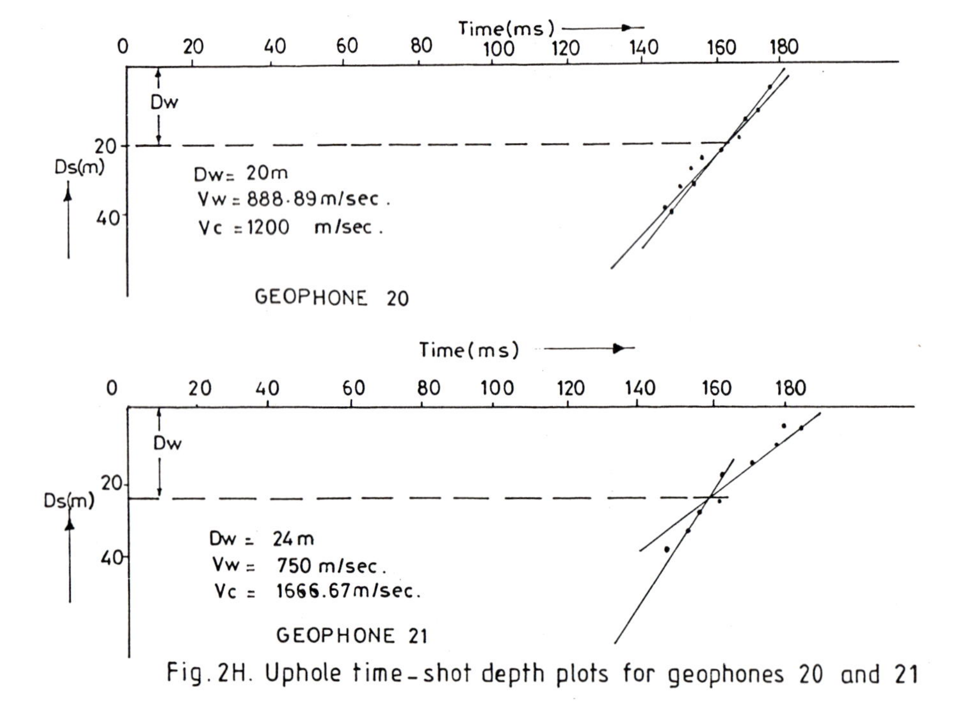

Fig. 2H. Uphole time-shot depth plots for geophones 20 and 21

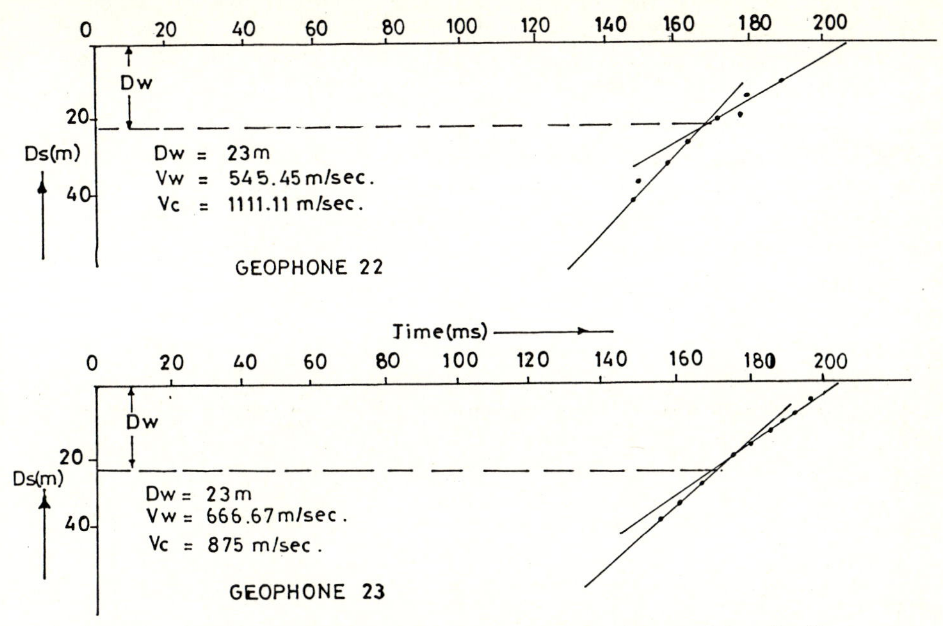

Fig.2I. Uphole time-shot depth plots for geophones 22 and 23

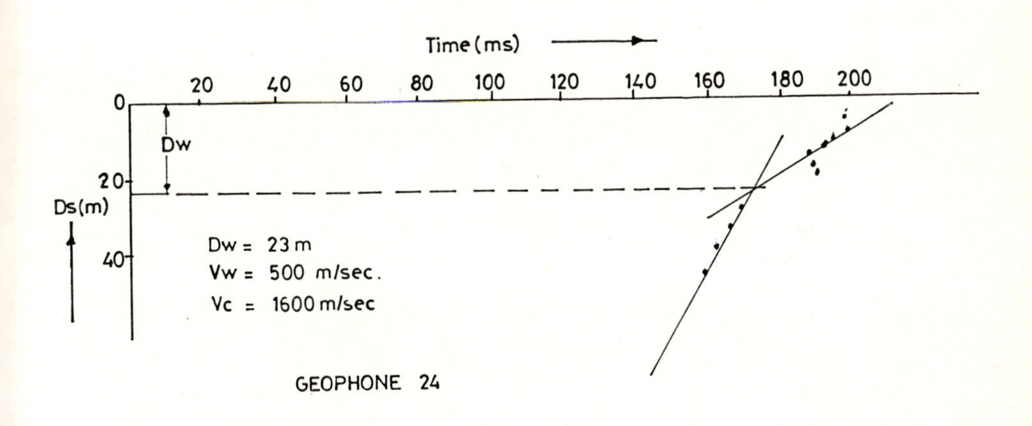

Fig. 2.J Uphole time-shot depth plots for geophone 24

RESULTS

For traces 1, 2, 3 (Figs. 2A, 2B, 2C) good results for the desired parameters were obtained. Beyond third trace/geophone (e.g. geophones 4, 5, 6, 7 and 8) reflected arrivals begin to occur and parabolic curves are obtained in the two layers (Figs. 2.A and 2.B) but Dw values still compare with those from earlier traces (Table 2). However, since reflected wave velocities vary with change in depth, neither Vc nor Vw could be obtained from these traces. Beyond trace 8 (i.e. 9-19) there seems to be an interplay of refracted and reflected one-way arrivals. Parabolic curves were obtained in the weathering layer and straight lines in the sub-weathering layer (Figs. 2.C - 2.G), but Dw values are unaffected. The highly anomalous velocities as obtained from traces 9, 10 and 12 (Fig. 2.C) can be attributed to a high velocity stratum below the geophones that received the signals or lateral variations within the weathered layer. Data from traces 20 to 24 (Figs. 2.H - 2.J) show that the geophones furthest away from the shot hole receive mainly refracted one-way arrivals, although interferring reflective one-way arrivals may occur (e.g. traces 21, 22 and 23 - Figs. 2.H and 2.I). Hence Dw, Vw and Vc have been determined fairly accurately.

In summary, data from this study suggest as follows:

a) Dw ranges from 16 to 24 metres with a regional average value of 20 metres;

b) Vw ranges from 430 to 600 m/sec., with a regional average value of 500 m/sec., while

c) Vc ranges from 1000 to 2500 m/sec., with a regional average values of 2000 m/sec. These values are within estimate for consodilated clays.

These values are not different from those obtained by the NNPC and Seismograph Service (Nig.) Ltd from Plus and Minus method of computation (N.N.P.C party 'X', Okeke, 1982).

Dw and Vw values represent the weathering parameters in the Onitsha area of unconsolidated sands (with limonite coating). Since all plots from this study suggest a two-layer situation, it is reasonable to assume from the geology (earlier discussed) that the sub-weathering layer that immediately underlies the weathered layer of soil and laterite is lignite seam or consolidated clay. We, therfore, believe that the depth of occurrence for lignite or consolidated clay in Onitsha - Oba axis is at least 16 metres.

DISCUSSION

Usually, where velocities are fairly constant during an uphole survey, straight lines are obtained when times are plotted against depths of shots. In normal areas, a single

Table 2 SUMMARY OF Dw, Vw and Vc for each Geophone

Geophone Number	Dw (Metres)	Vw (M/Sec.)	Vc (M/Sec.)
1	20.50	478.26	1666.67
2	23.00	434.78	250.00
3	20.50	444.44	1111.11
4	21.00	-	-
5	22.50	-	-
6	20.00	-	-
7	21.00	-	-
8	20.50	-	-
9	22.50	-	-
10	20.00	-	4333.33
11	23.50	-	2166.67
12	20.00	-	4333.33
13	20.00	-	2000.00
14	20.50	-	1666.67
15	21.00	-	1444.44
16	19.50	-	1250.00
17	18.00	-	833.33
18	16.00	-	1600.00
19	16.00	-	1000.00
20	20.00	888.89	1200.00
21	24.00	750.00	1666.67
22	23.00	545.45	1111.11
23	23.00	666.67	875.00
24	23.00	500.00	1600.00
Average or Regional Value	20.50	500.50	1859.26

high to low velocity break is normally shown at the base of weathering on all uphole geophones.

Arbitrary drawing of an average velocity line through a series of points will often lead to a serious misinterpretation. All information obtained should, therefore, be carefully studied before any anomalous point is disregarded. This is particularly true if an anomaly is repeated at all recording stations from a particular shot. If measurements of hole depth are exact, discrepancies have a sub-surface origin.

The main features identified in the uphole plots of this study are:

a) Single early arrivals with others falling on a straight line plot that may be due to floating charges,

b) Relatively high velocity stratum immediately above an observation point with a low-velocity interval beneath it,

c) Single late arrivals with all others falling on straight line. This may be due to error in depth measurement or hole fatigue. On the other hand, a late arrival may actually indicate a relatively low velocity stratum overlying a relatively high velocity layer, and

d) Changes in slope entirely due to velocity variations.

e) Vc data are so extreme that specific values cannot be satisfactorily quoted for lignite or consolidated clay as to be used in distinguishing between them. Continous Velocity logging (CVL) is recommened for such an exercise.

CONCLUSION

The results of this study suggest that the Dw can be estimated from all traces in an array of 24 geophones, including an uphole geophone. Vw has been best obtained from the first three and the last five geophones. Vc can be deduced from the first three geophones and a few of the intermediate ones between the fourth and nineteenth geophones. Over-all best results came from the first three geophones. Therefore, while we are aware that conventionally these parameters are usually calculated from information strictly obtained from the uphole geophone, this study suggests that since information from other geophones give fairly comparable values with the former, they should not be discarded. The Values of Dw, Vw and Vc obtained from this study are similar to those obtained with data from uphole geophone only and seismic refraction methods by major exploration companies in Nigeria for this region.

ACKNOWLEDGEMENT

We thank the N.N.P.C. Management for the permission to publish the data obtained from this study. We benefited from useful discussions with Professor S.O. Ifedili and Dr. C. Mgbatogu, both of the Physics Department, University of Benin, Benin City, Nigeria.

REFERENCES

Minerals and Industry in Nigeria, 1957: Federal Government of Nigeria Paper, Lagos.

Nigerian National Petroleum Corporation Party 'X', 1979: Unpublished Report.

Okeke, P.O., 1982: Thickness of lateritic materials around Onitsha, Nigeria. Nigerian Field, 47 (1), 104-107.

Seismograph Service (England) Ltd., 1978: Party 047 (unpublished Report).

Telford, W.M., L.P. Geldart, R.E. Sheriff and D.A. Keys, 1976: Applied Geophysics. Cambridge University Press.

Zoilkowski, A., 1982: Seismic Vital in Coal-Mining. Geophysics: The Leading Edge, 1, 33-35.

Distribution and Geology of Non-metallic Minerals in Nigeria

M. I. Odigi and C. O. Ofoegbu
Faculty of Science, University of Port Harcourt, P.M.B. 5323, Port Harcourt, Nigeria

Keywords: Non-metallic, Minerals, Basement complex, Sedimentary Basins

ABSTRACT

A study of the distribution, geology and genesis of non-metallic minerals in Nigeria has been carried out. Non-metallic minerals appear to be distributed in the major geological rock areas of Nigeria - sedimentary and basement complex rocks. Significant deposits of limestones, clays, glass, sand and gravels all of commercial importance are known to occur within the Cretaceous to Recent sedimentary environments while phosphates, gypsum, salts, nitrates, diatomites, fluospars and barytes are thought to occur in the Cretaceous - Tertiary sedimentary basins of Nigeria and are therefore presently being sought for and investigated. Occurrences of talc, asbestos, graphite, marble, dolomite and kaolin clay are associated with the schist belt of southwestern Nigeria, the hydrothermally altered and weathered rocks of the Younger Granites and the basement rocks of Nigeria. Other minerals which occur in uncommercial amounts in Nigeria include sillimanite, kyanite, nitrates, pumice and pumcite.

INTRODUCTION

Non-metallic minerals represent the third class of natural resources after the metallics and energy products. Most of the non-metallics in Nigeria are yet to be fully explored and exploited. The country imports most of her non-metallic mineral needs while the very few whose occurrence in Nigeria are well known have not been fully utilized. The systematic mapping and discovery of most of these minerals has been carried out by the Geological Survey of Nigeria although the Nigerian Mining Corporation (NMC) and the Nigerian Steel Council have in recent times embarked on the exploration for some non-metallics (industrial minerals) in different parts of the country. Kogbe and Obialo (1976) have presented the statistics of overall mineral production including some non-metallic minerals in Nigeria while Ford (1981) has discussed the observed association of economic minerals with the Benue Trough of Nigeria.

The present paper attempts to discuss aspects of the geology, distribution and genesis of the non-metallic mineral deposits of Nigeria as well as on-going exploration for some of these minerals.

GENERAL GEOLOGICAL SETTING

Non-metallic minerals are known to occur within three main geologic settings in Nigeria namely the crystalline rocks of the Basement Complex, Younger Granites and the sedimentary rocks. About more than half of the surface of Nigeria is covered by crystalline rocks and the other portion (less than 50 %) by sedimentary rocks. The crystalline Basement Complex rocks range in age from 2000 to 500 million years. They occur in three main areas, namely the northern, southwestern and southeastern Nigeria. Strips of younger sedimentary belts running along the Benue-Niger system more or less separates each of these basement blocks (Fig. 1) (Mc Curry, 1976; Oyawoye, 1972). Three major rock units can be recognised; the migmatite - gneissic complex rocks, Older Granite suites and ancient and younger metasediments.

The Basement Complex area is underlain by gneisses, migmatites and metasediments of Pre-Cambrian age which have been intruded by a series of granite rocks of late Pre-Cambrian to lower Paleozoic age (Mc Curry, 1976; Ofoegbu, 1985). The oldest rocks are represented by a series of Older Metasediments and gneisses believed to be of Birrimian age and Older. These rocks have been metamorphosed and granitised and the younger metasedimnets, are believed to have been deposited on this granitised basement during the Pan African Orogeny.

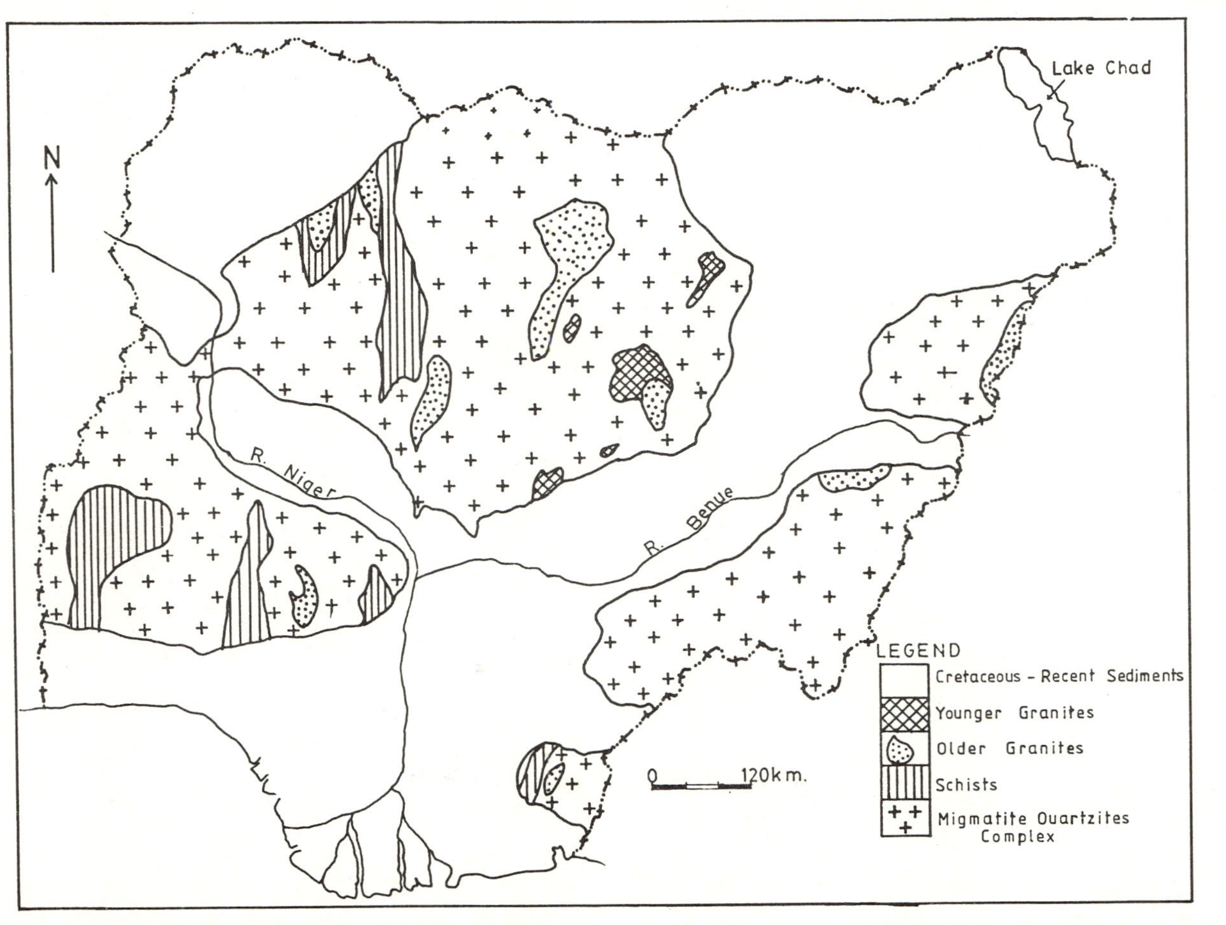

Figure 1. Simplified Geological Map of Nigeria.

Sedimentary Rocks of Nigeria

In the Pre-Cretaceous times, Nigeria consisted of uplifted continental land mass made up of Pre-Cambrian basement rocks which are unconformably overlain by Cretaceous sediments. The oldest known sedimentary sequence in Nigeria is the Albian to Santonian sediments and volcanics occupying a NE-SW trend in the Benue Trough (Fig. 1). The sedimentary rocks are up to 6000 m thick and comprise of sandstones, shale, limestone and evaporite pockets deposited in continental to marine environments. The earliest marine transgression occurred during the mid-Albian and was mainly confined to southeastern Nigeria and the Benue Trough (Fig. 2). The sedimentary basins witnessed series of trangressive and regressive phases which resulted in the deposition of both marine and deltaic sediments with their mineral deposits along the various depositional basins of southeastern, southwestern, northwestern, Niger, Chad and Benue Basins. The sediments suffered gentle folding, faulting and uplift during the Santonian to Campanian times.

After the end of tectonic event, the focus of sedimentation shifted to the south and west, with the deposition of shallow - water sediments during the Upper Cretaceous times in the southwestern, Niger, Sokoto and Anambra Basins. Lithologically, rocks of these basins are diverse, and comprise of basalt, sandstones, and congolomerates, ironstones, limestones, shales and coal measures in the Anambra Basin (Ofoegbu, 1985) (Fig. 2).

During the Tertiary times, sediments were confined to southwestern, northeastern Nigeria and the Niger Delta. In SW Nigeria in form of Ewekoro and Akinbo Formations. The sediments were mainly clay - shales, sandstones and limestones (Fig. 3a) while in the northwestern Nigeria, these deposits make up the Sokoto Group consisting of Dange, Kalambiana and Gamba Formations (Fig. 3b).

Post Maestrichtian folding movement affected different parts of the country. In the Benue valley all the sediments were folded about predominantly NE-SW axes with local bends in the E-W direction (Ofoegbu, 1985). At the later part of Tertiary period initiated the deposition of sediments of the Niger Delta where petroleum accumulated.

Volanic and Younger Granite Rocks

Magmatism affected Nigeria in Jurassic times and resulted in the emplacement of igneous and basaltic rocks. The Jurassic younger grantites of north central Nigeria are a suite of discordant high level, acidic to alkaline anaorogenic granites (Fig. 1). Their emplacement was accompanied by extensive acid volcanism, ring-faulting and cauldran subsidence (Mc Leod et al 1971). The major types of rocks are rhyolites, granites and porphyries with fayalite, aegirine - arferedsonite granites, biotite granites, syenites and trachytes, basaltic and basic rocks.

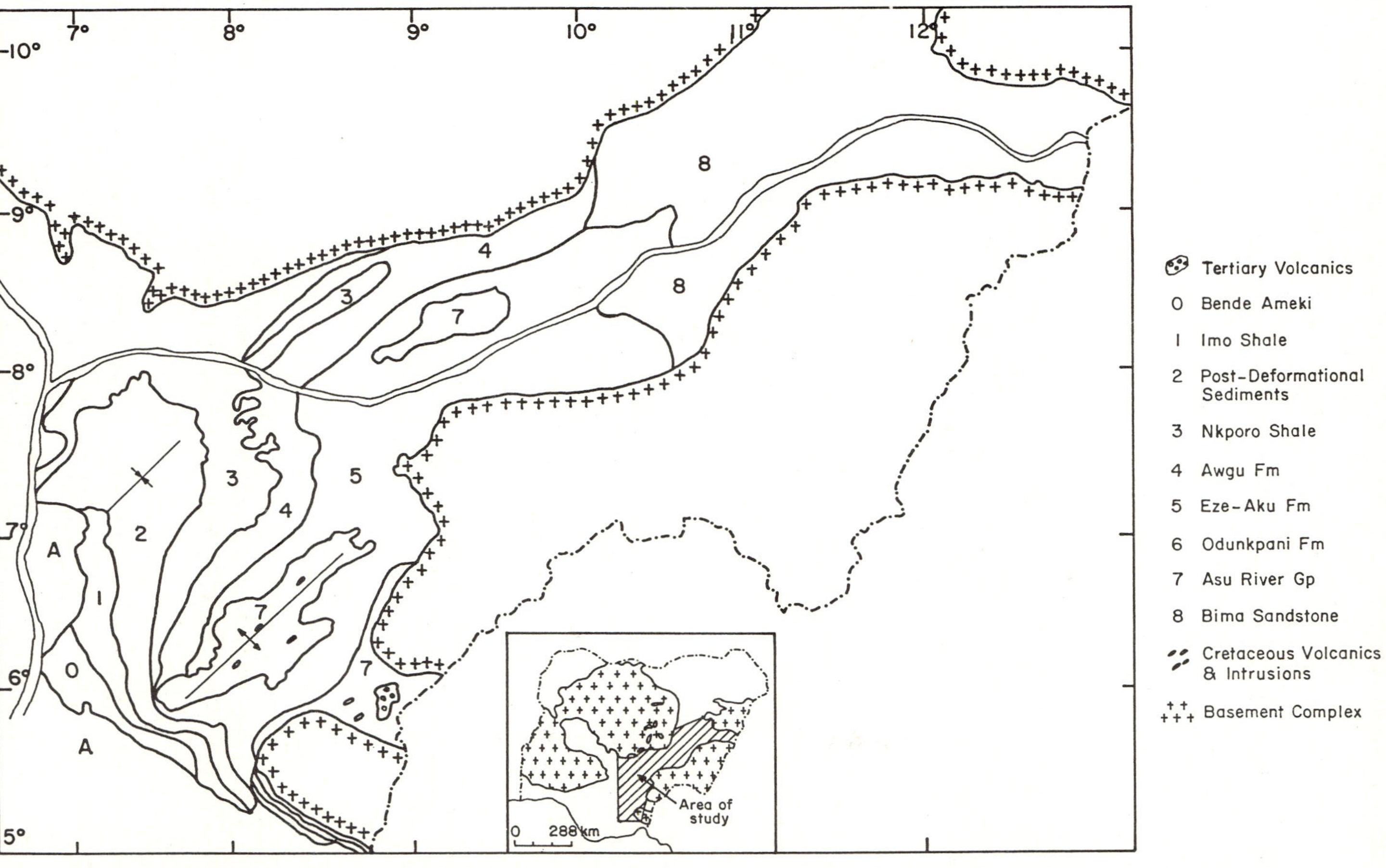

Figure 2. General Geology of Benue and Anambra Basins.

Magmatic activity occurred throughout Nigeria from Albian to late Cretaceous times but was confined mostly to the Benue Trough with the rock types ranging fron extrusive to intrusive. Volcanism occurred from Tertiary to most recent times in Nigeria. The effusives are dominantly basalts (older and newer), alkaline olivine basalts, phonolites and trachytes, and are confined mainly to northcentral and northeastern Nigeria as well as the Benue Trough.

DISTRIBUTION OF NON-METALLIC MINERALS

The distribution of non-metallic minerals in Nigeria, fit readily into the geologic pattern outlined above. They are here discussed under three general themes: those which reflect some kind of stratigraphic control and those related to granitoid plutonism and metamorphism.

Non-metallic Minerals of Sedimentary Type: Clay, Shale and Kaolin

The clay, shale and kaolin deposits of Nigeria can be grouped into two major types. The sedimentary argillaceous type (clay, mudstone and shale) and the sedimentary kaolin clay type which are transported clay rich from residual deposits. They usually form the alluvial or cacastrine clays. Significant deposits of clay/shale occur within the sedimentary basins of Nigeria. They are found within the Cretaceous, Tertiary to recent formations.

In the sedimentary basins of the Eastern Nigeria, huge quantities of clay, shale and kaolin deposits are widespread in the Asu River, Eze-Aku/Agwu Shale, Nkporo shale - Cretaceous Formations; and the Tertiary Imo and Ameki Formations (Fig. 2). Each of these formations have variable thickness and quality of clay, shale and kaolin deposits which vary from marine to continental clay, shale and kaolin deposits. Clay, shale and kaolin have been reported to occur at Ikpabo in Okigwe province with the clay possibly an intercalated member of the Ajali Sandstone and the Nsukka Formation. The clay deposit extends to Amuro about 4 km away and is not commonly exposed at the surface although the beds are almost horizontally disposed.

The over-burden thickness is considerably low. About 1.5 million tons have been proved to exist within a 200 acre plot in the area. Fire clay occurring below the No. 1 coal seam of the Mamu Formation at Enugu has been mapped for a distance of about 10 km (Ford, 1981). The thickness of the clay is estimated at 1.85 m on the average with a gradual decrease from north to south. The quantity available is estimated at 40 million cubic metres. The clays at Ikpabo and the Mamu Formation appear suitable for brick and several types of ceramic wares and are presently being utilised for these purposes. Sedimentary clays of Cretaceous age have been reported in the Okeluse, Ifon and Arimogija areas of Ondo State (Odigi, 1981). In eastern Nigeria, clay deposits are also distributed in the Tertiary formations and Benin Formation. One such deposit is the Itu Ibom clay which is of argillaceous material.

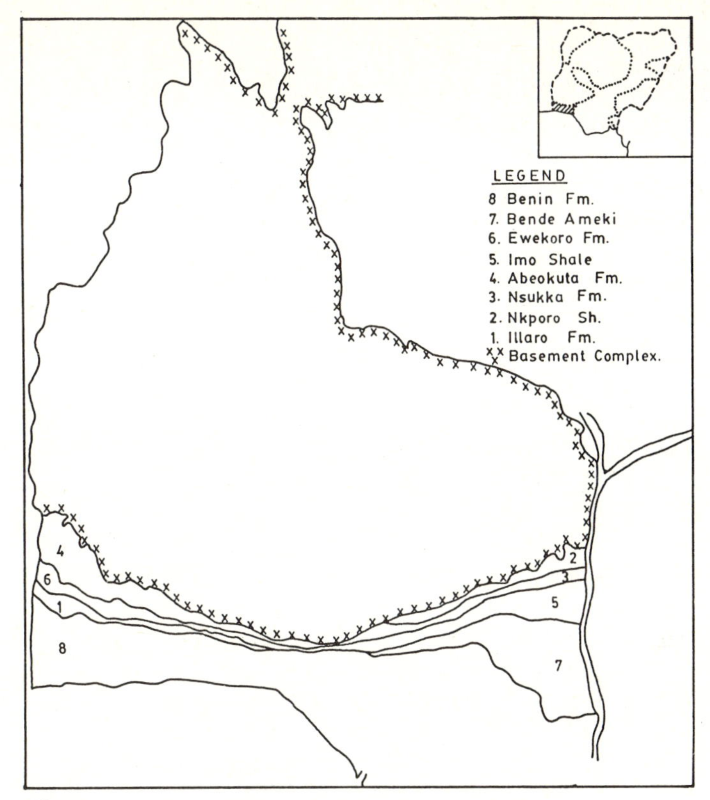

Figure 3a. General Geology of Dahomey Basin.

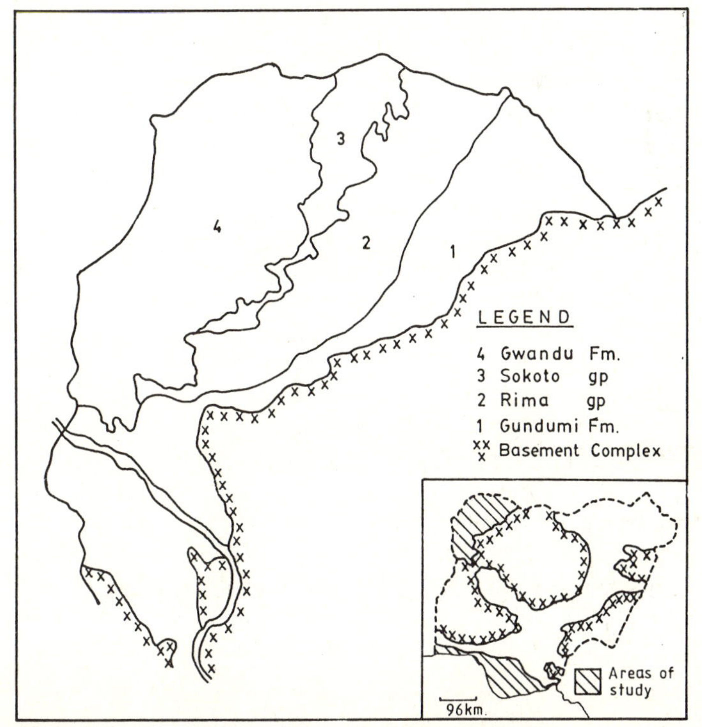

Figure 3b. General Geology of Lullemeden Basin.

The Tertiary Bende - Ameki and Imo Formations are also known to be rich in argillaceous matrials (Ogbukagu 1982).

In Ukwungu north of Issolu Ukwu, 3 mteres thick sedimentary kaolin is known to occur in the Bende - Ameki Formation. The Ogwashi - Asaba Formation is characterised by its lignite content with an association of dark clay horizon. The formation is restricted to midwestern and eastern Nigerian sedimentary basins. Sedimentary kaolin clays also occur in the Benin Formation as thick lenses overlain by thick overburden. Huge deposits of sedimentary kaolin are exposed along the river gullies around Siluko and Oza - Nogoro in Bendel State. Elsewhere in some parts of southeastern Nigeria, large argillaceous sediments of mudstone and grey kaoline occur within the Benin Formation. They occupy large areas of Itu, Ikomo and Ibom (Odigi, 1986).

Clay deposits are well distributed in the Chad Formation in northwestern Nigeria which varies lithologically both laterally and vertically. Towards the centre of the basin, lacustrine clays are predominant in the sequence but near the margins fluvialite sand, grits and gravels become more important. Within the Nigerian portion of the basin, it is essentially an argillaceous sequence and sandy layers are separated by clay deposits around Maiduguri (Matheis, 1976).

The Illummeden Basin of northwestern Nigeria houses argillaceous sediments of mainly Cretaceous - Tertiary age. The argillites whose thickness varies from place to place occur in the Illo-Gundumi Formation, consisting of grits and clays overlain unconformably by Upper Cretaceous Taloka and Wurno Formations which are made up of mudstones and sandstones respectively (Kogbe, 1976). Good sedimentary clays occur at Ala-Igbite and Itori in the Cretaceous formations. These are believed to predate the Cretaceous sediments. Some of the Tertiary sedimentary structures of southwestern Nigeria are also characterised by the occurrence of clay and shale. The Ewekoro Formation, though, dominantly limestone has an alteration of shale layers and is overlain by the Akinbo Formation consisting of argillaceous sediments of shale, mudstone, silt and limestone bands. The clay-shale deposits of the Akinbo Formation are presently being exploited and used in the petroleum, cement, brick and paper industries (Odukoya, 1981). The Oshosun Formation consists of about 18 m thick clay (GSN Borehole No. 1582) while the Ilaro Formation consists of a 4 m thick gritty clay (Kogbe, 1976b). Quaternary clay deposits have also been reported in the Niger Delta Swamps, along the valleys of streams, rivers and flood plains where they occur as alluvial deposits. Notable occurrences are those at Iguoriakhi - Benin West, Eket and Etinam in Cross River State (Fig. 4).

Limestone and Dolomite Deposits

Limestone is one of the commonest non-metallic minerals found in Nigeria. Its occurence is wide spread in almost all sedimentary basins of Nigeria. The limestones occur principally in the Albian - Paleocene formations and are of good quality. Early

Cretaceous limestone rocks whose thickness average 10 - 15 m occur around Yandev which is 8 km east of Gboko. The limestone which occur mainly in the Albian formation are of good quality and are used for cement making at Yandev. The most important outcrop is the Yandev Limestone while other outcrops are found in the area south of Makurdi, Arufu, Giza and Keana areas. Upper Cretaceous limestones which occur at Mfamosing and Arochukwu belong to the Upper Cretaceous Odukupani Formation. The limestone deposits of Mfamosing are important as they are exploited for the cement factory at Calabar and as a raw material for the Aladja Steel Factory while the extensive deposits reported at Arochukwu are yet to find full economic use. Other Upper Cretaceous limestone deposits are those that occur at Nkalagu (Peters and Ekweozor, 1982) and Pindiga and Gongola Formations (Reyment, 1965). These are presently being quarried for the cement factories, at Nkalagu and Ashaka respectively.

Limestone deposits of Lower Tertiary (Paleocene) age occur in Okeluse, Ewekoro and Sokoto within the clayey, limestone and shale formations of Ewekoro and Sokoto Group sediments respectively. The Okeluse limestone is believed to be a lateral eastern out-crop of the Ewekoro Formation and the deposit is mainly a sequence of blue-grey clayey shales with occasional thin bands of marls and limestone. The Imo Shale passes laterally into a thick shelly, arenaceous limestone (Ewekoro Formation) of about 10 metres thick. The Ewekoro limestone is mined for cement at Shagamu. The Kalambaina Formation (Sokoto Group) is another source of good quality limestone mined for cement making. This formation consists mainly of limestone and shale sequence. Good exposures of the formation are scarce except in the quarry of the cement factory and in well sections. The type locality is at Kalambaina and a type section from the quarry has been described by Kogbe (1973). The maximum thickness of the formation recorded from boreholes is below 20 metres (Kogbe, 1976b).

Limestone occurrences at Nkalagu, Ashaka, Yandev and Awe attain an average thickness of about 6 m and contain abundant fossil amonites. Futher outcrops of thinner beds occur at Makurdi, Ruma River bank at Ribi and New Zurak. Ford (1981) reported the occurrence of thicker outcrops of limestones in the area of the Upper Benue, most of them between 10 and 50 cm thick. The best sedimentary limestones in the country therefore appears to be found in the Benue Trough but these require careful and detail mapping to ascertain their stratigraphy and lateral extensions.

There are no known significant occurrence of sedimentary dolomite outcrops in Nigeria, except cases of dolomitisation in the vicinty of limestone beds along the Calabar Flank and Ewekoro (Peters, 1978).

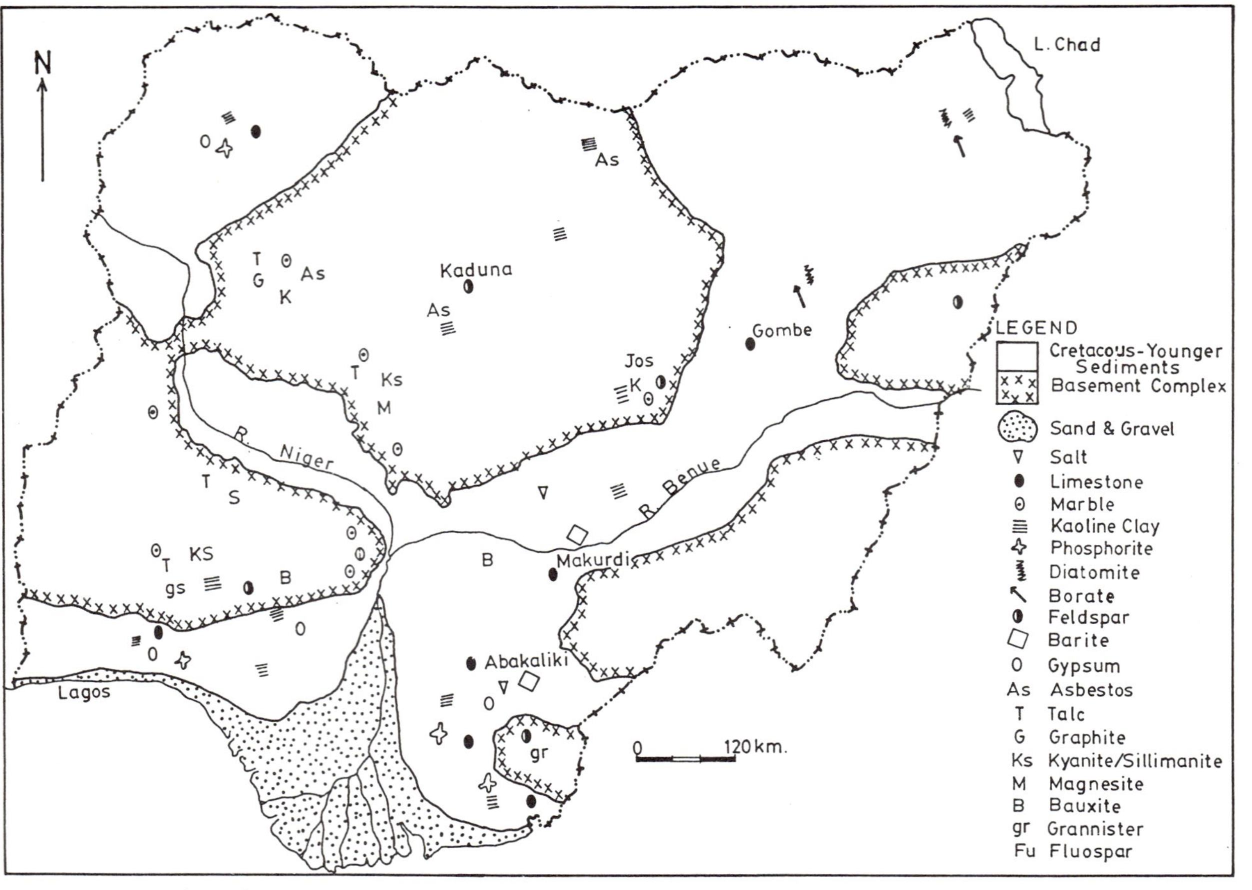

Figure 4. Location of Non-metallic Minerals in Nigeria.

Glass Sand, Phosphate, Salt and Gypsum Deposits.

Silica sands occur in beaches, lagoons and within the coastal plain environments stretching from Dahomey, Okitukpa, Niger Delta to Cameroons (Fig. 4). Massive white quartz sandstone outcrops have been reported (Ford, 1981) at Bashar in the Middle Benue and it is generally believed that extension and occurrence of similar sandstone bodies beyond the confines of the area is possible within the Cretaceous sedimentary rocks. It is also reported that there are other local segregations of relatively pure quartz among the very widespread sandstone outcrops of the Middle and Upper Benue Trough.

Some silica sands are known to occur along the big rivers of Imo, Forcados, Qua Ibo, Bonny, Ogun and others that empty into the sea, but contain high amount of impurities like iron materials, shell fragments and light and heavy mineral concentrates. Beach and lagoonal sands are noted to be free of iron staining and nearly white in colour. they also contain lime in form of shell fragments. The quartz sandstones are friable while the beach and lagoonal sands are unconsolidated. Glass factories in Nigeria quarry glass sands from lagoons, beaches and from these rivers.

Glass sands used in the Glass Industries occur in the Tertiary to Recent sediments. The occurence of glass sand near Bende-Imo State was investigated and found to be uneconomic. However, an extension of the sands beyond this area is probable.

Phosphatic beds are known to occur in Tertiary sedimentary rocks and are of biochemical sedimentary type. Phosphatic sediments have been recorded in the surface and subsurface sediments of the Eocene - Oshosun, Illaro, Ameki and Dukameja Formations respectively (Reyment, 1965). The only well verified significant occurrence of phosphate is however the Eocene sediments which contain 3 m thick phosphate sediments (Oshosun and Illaro). The main outcrop of the phosphatic beds occur within an area bounded by Oshosun, Afo Ajio and the 49 km post of the railway line. The formation consist of laminated clay and phosphate rich in glauconite, fish teeth and animal microfossils. Phosphate sediments also occur in a borehole at Ijoko southwestern Nigeria. Preliminary prospecting and exploration activities has been carried out by the Geological Survey of Nigeria to assess the stratigraphy, lateral and vertical extensions within the Oshosun, Illaro and Ameki Formations which were deposited in a similar environment. The phosphate beds if proved to be economical would supply raw materials to the Fertilizer Industry at Onne, Rivers State. A systematic study of the Eocene sediments is therefore needed to reveal the extent of the phosphate deposits.

Evaporite sequences are common in the sedimentary basins of Nigeria (Fig. 4). They are found in Upper Cretaceous and Tertiary sediments. The presence of evaporites has been known for a long time in the country. Attempts to explore the salt deposits especially in the Benue Trough have been initiated by the Nigerian Mining Corporation while the gypsum deposits are yet to be investigated due to the fact that the known deposits of gypsum invariably occur in disseminated small crystalline masses and irregular veinlets and are usually overlain by thick overburden. Gypsum is known to occur in the areas of Egoli and Orbiokhuan in Palaeoene - Lower Eocene sediments of Imo Clay-Shale Formation and at Iwutu near Afikpo in the Cretaceous sediments. Egoli lies 8 km south of Auchi while Orbiokhuan is 2 km north of Sabongida Ora (Bendel State). The mineral is exposed in two thin discontinuous horizons at a road side cutting, crystallised within the bedding of the shales and as thick crack fillings. The thickness of the gypsiferous section varies from 3 cm to 10 cm in bedding planes and less in the cracks, separated by layers of shale. There are numerous salt water springs along the entire length of the Benue Trough. The springs are frequently associated with fractures and flow freely at the surface, often giving rise to salt pools which form important local sources of salt. There are also salt ponds in the vicinity of Uburu and Okposi (Ford 1981).

The salt and gypsum deposits are usually interbedded with shales and limestone units. All the formations in association with these deposits were deposited in a marine condition under restricted and confined basin environment. Restricted connection with the sea gave rise to evaporitic conditions in the basin which prevailed for most of the depositional stage. Carbonate buildup in the basins may have led to the restricted conditions which prevailed only partly into the Upper Cretaceous and Tertiary times, before re-adjustment in the basin led to a major marine transgression. Shale interbeds and beds of shaly gypsum are therefore common, and volumetrically important in places. Little or no chemical analyses are available, thus little can be said about the salt and gypsum deposits. Torna has also been reported from a salt water spring near Keana. The present indications are that the Torna is unlikely to be considered for economic development.

Other Non-Metallic Minerals of Sedimentary Type in Nigeria

In addition to the non-metalllics discussed in the preceeding sections are the Barates, Nitrates, Diatomites and Barites, also known to occur within the sedimentary rocks of Nigeria. Barates and Nitrates have been reported to occur in the Benue Trough (Fig. 4) as brines of saline lakes, saline springs and in the Chad Basin. Exploration work is being carried out along the Middle Benue Trough to determine the locations and extent of these lake deposits.

Diatomite occurs in the Chad Basin and north end of the Upper Benue Trough in the form of lake deposits. Ford (1981) reported the lake deposits to be in a small depression on the surface of the Cretaceous sediments at Bularaba (Fig. 4). The deposits, scarcely up to 3 m in thickness, have been estimated to contain 55,000 tons of good quality diatomite. The diatomites are of Pleistocene age and interbedded in the Chad Formation. Other occurrences of diatomite between Akabire and Gujba have been tested for their commercial value and prove to be good as factory materials.

Barite an important ingredient for the manufacture of drilling mud has been reported to occur in the Middle Benue Valley (Ford, 1981). The mineral occurs in several hydrothermal veins and lobes in the Benue Valley (Fig. 4) at Aloshi, Akiri - Wase, Azara, Ibi, Keana, Chiata and Gbande near Gboko. Most of these occurrences are associated with the Cretaceous sandstones. The mineral appears to infill fractures and in most cases associated with Pb-Zn mineralisation zones. Also these barites occur with quartz, siderite and arikerite the carbonates of iron, calcium and magnesium.

NON-METALLIC MINERAL DEPOSITS OF IGNEOUS TYPE

Classified under these are non-metallic minerals derived from both igneous activity and the alteration of igneous rocks. Deposits of this type include feldspars, hydrothermal volcanic clays, bauxitic clays, flouspar and aggregates. Feldspars occur in the granite pegmatites which are found in granitic intrusions, some around high grade metamorphic rocks/pegmatites containing the commercial variety of feldspar e.g potash and soda are found in the Older Granite rocks around Okene, Ijero Ekiti, between Kaduna and Zaria, 6 km south of Dengi and in high grade metamorphic areas of the Oban massif where they occur as small ridges (Fig. 4). Occurrence are locally extensive as in the case of Ijero Ekiti but in general have not been mapped. Some studies have, however, been made on the material in certain localities, for example, the Liga Hill pegamites south of Maiduguri which was found to be about 100 m long, 8 m wide and made up of potash feldspar (Okeke and Fitton, 1982). The feldspar deposits may be of interest to the ceramic industry.

The Cretaceous and Tertiary volcanic activites in Nigeria is an indirect evidence to suggest the existance of volcanic clays (primary and secondary) in the crystalline/igneous terranes in Nigeria. Primary kaolin clay deposits have been recognised in several localities in the Nigerian Basement area (Fig. 4). These deposits are weathered products from granites, gneisses, pegmatites and schists. Three distinct geologic areas were affected by hydrothermal alteration and weathering process: Younger Granites (Jos, Rayfield, Bukuru, Gana Wuri), undifferentiated Older Granites (Omi Adio, Iperindo and Akinlobi) (Ajayi and Agagu, 1981) and contact areas (Fig. 4) between the basement and the Cretaceous sediments at Ifon (Odigi, 1981). High quality kaolinitic clays occur at Kankara, north of Malumfashi, Zaria Province and are believed to have been derived from the alteration of basement gneisses which form the surrounding country rocks. Their shape and the presence of vermiculite veins and secondary chalcedomic silica, is consistent with an origin by hydrothermal alteration from an underlying source, presumably magmatic (McCurry, 1970). Recent borehole investigations of the deposits have shown that highly altered rocks of volcanic aspect underlie the clays at around 50 m depth and hydrothermal emanations could be responsible for kaolination of the basement rocks. The occurrence of bauxitic clay in Nigeria is relatively few but there are showings on the altered basaltic lavas of the Jos Plateau having a thickness of 30 m capped by a lateritic iron stone layer, and at Orin Ekiti

(Ondo State) and Oturkpo (Benue State), which are products of residual weathering of basement and aluminous sedimentary rocks respectively although none have been proved to be economic in extent. Residual clays derived from the weathering of granitic and metamorphic rocks of the Basement Complex also occur in Nigeria. Sizeable deposits of these occur at Igbetti, Omi-Adio and Omialafara clays are weathered products of marble and gneissic rocks respectively while that at Akinlobi is a product of pegmatitic weathering. Flouspar occur in veins with galena at Arufu and Akwana in the Middle Benue Valley and as wall-rocks of limestone that have been silicified. The extent and quality of these occurrences require detail assessment. A large proportion of the land surface of Nigeria is covered with various rock types including unconsolidated material which are major sources of aggregates used as construction materials. Aggregate production in Nigeria will be divided into production from granular deposits and from stone quarry. Generally, the areas underlain by granitic and some metamorphic rocks produce good to excellent aggregate material while most of the sedimentary rocks do not produce concrete or asphalt - quality aggregate. Within the Benue Trough, metamorphic rocks are restricted to small areas such as Gombe and Kaltungo inliers. Elsewhere stone aggregates, are the basalts, gabbro and dolerites of the volcanoes and igneous intrusions. These igneous rocks are widespread in the Benue Trough and range in age from Cretaceous to Recent. Most of the sand and gravel deposits are located along river valleys and coastal areas, (Fig. 4) which are also considered suitable for agriculture, residential and recreational development. The Jurassic, Cretaceous and Tertiary volcanics of Nigeria are possible sources of pumice and pumicite materials. They occur at small localities south of Langtang (Jos Plateau) (Fig. 4) (Odigi, 1976) but these outcrops have not been mapped in any detail to assess the quantity.

NON-METALLIC MINERALS DEPOSITS OF METAMORPHIC TYPE

Several non-metallic minerals occur within the metamorphic terrains of Nigeria. These include marble, dolomite, asbestos, talc, kyanite, sillimanite, magnesite and quartzite. Marble deposits in Nigeria are often associated with metasediments such as schists, amphibolite complexes, meta-conglomerates and calc-gneisses all of Pre-Cambrian age.

Marble occurrences (Fig. 4) of commercial quantity are being exploited at Jukura, Igbetti and Ukpilla while smaller occurrences near the Ogbo River were reported by Jones and Hockey (1964). The Marbles are of varying colours of white, greyish, cream and pale greenish types and composed of calcite with minor amounts of graphite and calc-silicate minerals. Marble is also found interbedded with gneisses and slightly migmatised schist near Amper north of sheet 191 of the Geological Map of Nigeria. Kwakuti Marble is interbedded with quarzite and schist in foliated gneiss, in the village of Kwakuti. Most samples contain dolomite and magnesite in addition to calcite, tremolite and antigorite. In Nigeria today marble is being worked as a material for building, decoration and monument.

Dolomite has been reported around Osara and Burum within the Pre-Cambrian basemant of Kwara State and New Federal Capital area respectively (Fig. 4) (Scott et al., 1983). The Burum dolomite is still under investigation, while the Osara dolomite deposit was found to be small. These could be sources of possible raw material for dolomite refactory manufacture and a flux material for the Steel Plant at Ajaokuta. Asbestos on the other hand has been reported in the amphibolite-schist of northwestern Nigeria. The amphibole variety is believed to occur in a schist terrane. It is suggested that asbestos could be prospected for in the amphibolite schist, ultra basic and limestone or dolomite rocks that are altered. Asbestos is a commercially important raw material in the manufacture of asbestos cement and insultors which at the moment are imported into Nigeria.

Talc is found in the northern part of the Federal Capital Territory. The extent is yet to be mapped out but it is in association with metasediments and intrusions. The occurrence of talc has been reported in sheet 31 of Geological Map of Nigeria often associated with chlorite schists. They are grey in colour and composed principally of talc with varying percentages of carbonate.

Kyanite and sillimanite occur in the Nigerian Basement Complex. Geological investigations have shown that at Tunga Bargwoma (Birin Gwari) and west of Dengi (Wase), kyanite occurs as layers and segregations within schist and quartzite. Sillimanite has also been found as an accessory component in quartzites at several localities at Erin-Oke (Oyo State) and Omuaran (Kwara State), but none of these are of commercial significance (Fig. 4). Kyanite and sillimanite are raw materials for refractories used in steel making.

Magnesite is usually found in carbonate or as alteration product of ultramafic rocks. Magnesite occurrence in Nigeria is localised within the Younger Metasediments as ultramafic schist. Zones of talc magnesite have been mapped at some localities in the New Federal Capital Territory area. Magnesite is also used as refractories. Graphite mineral although not reported to occur in large amounts has been found in the metasedimentary terranes of Nigeria. Detailed mapping of this area with metamorphosed carbonaceous sediments is likely to reveal areas with commercial amount or showing of graphites.

Quarzite usually known as gannister is the main raw material for silica refractory when pure. Quartzite is abundant in the metamorphic terrane such as in the Basement Complex of Nigeria. In Nigeria, there are two groups of quartzites; the group that are found within the 'Older Metasediments' of the migmatite-complex and the 'Younger Metasediments' of the schist belts. Quartzites for silica refractory are found within the Younger Metasediments; for example, the quarzite outcrops of the Effon Psammite Formation, at Effon Alaye and Ijero in Oyo State (Fig. 4). Geological investigations have also shown that low grade quartzite occur in the metamorphic areas of the Oban massif Cross Rivers State (Odigi, 1986).

DISCUSSION AND CONCLUSION

Although only a few economic non-metallic or industrial minerals are mined in Nigeria there is considerable potential for finding new areas and extension of the use of these materials. From the geologic point of view, the following areas in Nigeria have been mapped out as potential regions for the occurrence of non-metallic minerals.

1. The Chad Basin (Tertiary-Recent) which contains diatomites, nitrates, refractory clays and glass sands.

2. The Sokoto Basin (Cretaceous-Tertiary) in northwestern Nigeria which possesses large deposits of limestones, clays and thin layers of gypsum interbedded with the limestones. The presently known deposits of gypsum and phosphate requires exploration while regional and detail mapping and exploration work is needed to cover the basin for possible occurrence of other non-metallic minerals.

3. In the Dahomaeyan Basin in southwestern Nigeria, large deposits of limestones, clays and glass sand occur. These materials serve as raw materials for various industries such as cement, glass (bottle) manufacturing and ceramic industries. Recent geological mapping and exploration program by the Geological Survey of Nigeria has discovered the occurrence of phosphates in the basin. Lenses of gypsum are found in association with limestones and a drilling program should be initiated to assess the extent of subsurface occurrence of these.

4. The Anambra Basin in the southeastern Nigeria, has potential for argillaceous sediments that could be evaluated for industrial uses and also lenses of gypsum are reported to occur in the area. The basin holds high prospects for phosphate and glass sands. There is an extensive limestone deposit from the Calabar Flank to Arochukwu where the limestones are dolomitized in places and it is believed that this belt of limestone extends into Nkalagu Formation (Peters and Ekweozor, 1982).

5. The Benue Trough, filled with Cretaceous sediments has potential for limestones, clays, glass sand, fluospar, barite and salts.

   The known deposits of barite, salts and fluospar are small but require further mapping and exploration program. Recently, exploration work has been initiated by the Nigerian Mining Corporation to determine the potentials of salts in the Middle Benue Trough. It is believed that a similar program is capable of yielding useful data on bentonitic clay, fluospar and barite.

6. The Pre-Cambrian Schist belts and the Older Granite rocks of Nigeria posses various potentials of non-metallic minerals such as marble, feldspar, magnesite, talc, asbestos, graphite and sillimanite/kyanite. The basement complex province provide a varied environment for the formation of these non-metallics. Marble and feldspar are commercially mined while magnesite, talc, asbestos, graphite and kyanite/sillimanite have been reported. These minerals merit further exploration while extensive chemical studies of talc deposits should be encouraged.

7. The Felsic and alkaline ring-dyke granitic complexes of the Younger Granite province have indications for pumice and pumcite materials and deposits of kaoline clay. The province posses geologic characteristics similar to known and mineralised Younger Granites, for example Cornwall for its kaolin deposits. Considerable potential exists for the discovery of deposits of kaolin in the Younger Granite province.

The geologic environments as discussed above holds considerable potential of non-metallic or industrial minerals of economic value, in addition to those already known. Large parts of the area are unexplored as far as the geology and mineral resources are concerned. Even those parts which have been mapped may contain near surface deposits. Field geology, geophysical and drilling work on promising areas may reveal greater deposits than we presently imagine. Deposits that occur in smaller areas and probably disseminated minerals can be extracted at a profit. Some of the Cretaceous to Tertiary argillaceous rocks require evaluation to assess their non-metallic mineral potentials.

Non-metallic minerals and rocks have been the bedrock of the nation's past and present industralization. Evidence do exist in the past archaelogical records that considerable use of non-metallic materials supported local makers of pottery, brickwork, ceramics, iron and bronze artifacts of the Nuk and Ife-Benin Kingdoms.

With the new and dynamic industrialization policy in which industries have to be established, the need for proper evaluation of the present occurrences be it major, minor or traces should be embarked upon by Federal and State Governments and interested private Industrialists. For Nigeria to emerge stronger in her industralization goals, as well as conserve her foreign exchange, an inventory and development of these resources is needed urgently. Most major structural components utilise non-metallic minerals and rocks in industries such as the Iron and Steel, Cement, Ceramics, Paint, Plastic and Rubber Glass to mention a few. If these industries and some new industries that are yet to be established are to be self-reliant and to improve on the economy, then efforts to encourage Research Institutes and Universities are needed.

A non-metallic Resource Inventory is an essential program in a developing country such as Nigeria and should be initiated to provide basic information towards sound planning and management of industrial or non-metallic materials. This would also show the quality, present degree of use and potentials of the deposits.

REFERENCES

Ajayi, J.O. and O.K. Agagu, 1981: Mineralogy of primary clay deposits in the Basement Complex of Nigeria. Nig. J. Min. Geol., 18, 27-30.

Ford, S.O:, 1981: The economic mineral resources of the Benue Trough. Earth Evol. Sciences, 1, 154-163.

Jones, H.A. and R.D. Hockey, 1964: The geology of Southern Nigeria. Geological Survey Nigeria Bull. No. 31.

Kogbe, C.A., 1973: Geology of the Upper Cretaceous and Tertiary Sediments of the Nigerian Sector of the Lullemeden basin (West Africa). Geol. Rundsch., 62, 197-211.

Kogbe, C.A., 1976a: The Cretaceous and Paleocene sediments of Southern Nigeria. In: Geology of Nigeria (C.A. Kogbe, Ed.), 273-282. Elizabethan Press, Lagos.

Kogbe, C.A., 1976b: Outline of the geology of the Lullemeden Basin (N.W. Nigeria). In: Geology of Nigeria (C.A. Kogbe, Ed.), 331-338. Elizabethan Press, Lagos.

Kogbe, C.A. and A.U. Obialo, 1976: Statistics of mineral production in Nigeria (1946-1974) and the contribution of the mineral industry to the Nigerian economy. In: Geology of Nigeria. (C.A. Kogbe, Ed.), 391-428. Elizabethan Press, Lagos.

Matheis, G., 1976: Short review of the geology of the Chad Basin. In: Geology of Nigeria (C.A. Kogbe, Ed.), 289-294. Elizabethan Press, Lagos.

McLeod, W.N., D.C. Turner and E.P. Wright, 1971: The geology of Jos Plateau (Vol. 1). Geol. Survey Nigeria, Bull. No. 32.

McCurry, P., 1970: The geology of degree sheet 21 (Zaria). Unpublished M.Sc. thesis, Ahmadu Bello University, Zaria.

McCurry, P., 1976: The geology of the Pre-Cambrian to lower Palaeozoic rocks in Northern Nigeria - a review. In: Geology of Nigeria (C.A. Kogbe, Ed.), 15-39. Elizabethan Press, Lagos.

Odigi, M.I., 1976: Annual Report of Geological Survey of Nigeria.

Odigi, M.I., 1981: Appraisal of clay deposits at Ifon and its environs. Unpublished M.Sc. thesis, University of Ibadan, Ibadan.

Odigi, M.I., 1986: Clay resources of some parts of S.E. Nigeria. Proceedings of Workshop on clays (In Press).

Oduakoya, A.A., 1981: Mineralogy, physical and chemical properties of the clay-shale above the Ewekoro Limestone in S.W. Nigeria. Unpublished M.Sc. thesis, University of Ibadan, Ibadan.

Ofoegbu, C.O., 1985: A review of the geology of the Benue Trough of Nigeria. J. Afr. Earth Sci., 3, 283-291.

Ogbukagu, K.N., 1982: Properties and benefication of argillaceous rocks of Southern Nigeria sedimentary basins. Nig. J. Min. Geol., 19, 43-51.

Okeke, P.I. and J.G. Fitton, 1982: Liga Hill pegmatites. Paper presented at the 18th Annual Conference of Nigerian Mining and Geosciences Society.

Oti, M.N., 1983: Petrology, diagenesis and phosphate - mineralization in Cretaceous limestone in Arochukwu/Ohafia area, South-East, Nigeria. Nig. J. Min. Geol. 20, 95-104.

Oyawoye, M.O., 1972: The basement complex of Nigeria. In: African Geology (T.F.J. Dessauvagie and A.J. Whiteman, Eds.), 67-99. Ibadan University Press, Ibadan.

Peters, S.W., 1978: Dolomitization of the Ewekoro Limestone, Nig. J. Min. Geol., 15, 79-84.

Peters, S.W. and C.M. Ekweozor, 1982: Petroleum geology of the Benue Trough and S.E. Chad Basin, Nigeria. AAPG Bull., 66, 1141-1149.

Reyment, R.A., 1965: Aspects of The Geology of Nigeria. Ibadan University Press, Ibadan.

Scott, P.W., T.A.R. Thanoon and C.O.F. Arodiogbu, 1983: Evaluation of limestone and dolomite deposits. Proceedings of Extractive Industry Geology Conference, 107-126.

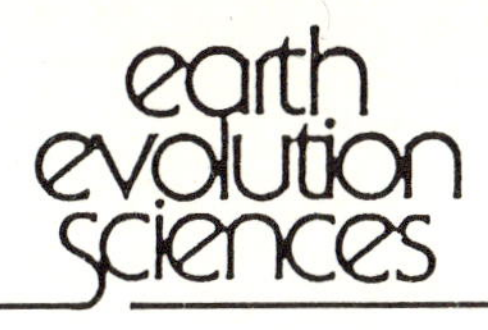

International Monographic Series
on Interdisciplinary
Earth Science Research and Applications

Editor
Andreas Vogel, Berlin

Volumes available:

Rodney A Gayer (Ed.)
The Tectonic Evolution of the Caledonide-Appalachian Orogen

Andreas Vogel
Hubert Miller
Reinhard O. Greiling (Eds.)
The Renish Massif

Jean Pohl (Ed.)
Research in Terrestrial Impact Structures

Chi-Yu King
Roberto Scarpa (Eds.)
Modeling of Volcanic Processes

Samir El-Gaby
Reinhard O. Greiling (Eds.)
The Pan-African Belt of Northeast Africa and Adjacent Areas

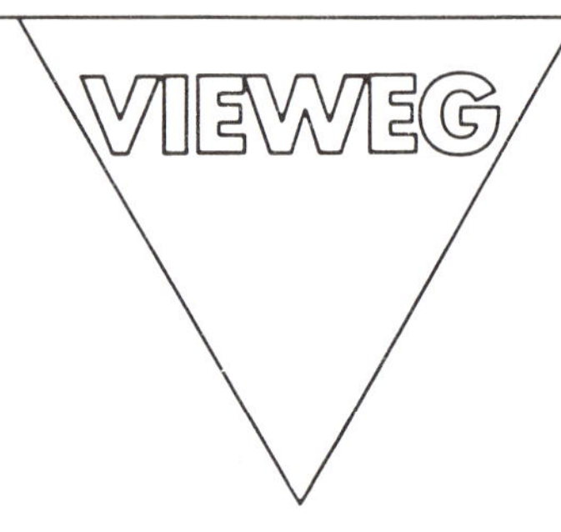

Theory and Practice of Applied Geophysics

comprising volumes with collections of papers on the geophysical exploration of the earth's interior for scientific purposes as well as practical applications in prospection for minerals deposits, exploration of ground water and the solution of problems in engineering geology.

Volume 1

Andreas Vogel (Ed.)
Model Optimization in Exploration Geophysics
Proceedings of the 4th International Mathematical Geophysics Seminar held at the Free University of Berlin, February 6 – 8, 1986.
Editorial Advisers: Rudolf Gorenflo and Gerhard Berendt. 1987.
VIII, 396 pp. 17 x 24,5 cm. Hardcover

Volume 2

Andreas Vogel (Ed.)
Model Optimization in Exploration Geophysics, Part. 2
Proceedings of the 5th International Mathematical Geophysics Seminar held at the Free University of Berlin, February 4 – 7, 1987.
1988. Approx. 380 pp. 16,5 x 24 cm. Hardcover